THE *Field Guide* TO CHICKENS

THE *Field Guide* TO CHICKENS

By Pam Percy

Voyageur Press

First published in 2006 by MBI Publishing Company and Voyageur Press, an imprint of MBI Publishing Company, 400 1st Avenue North, Suite 300, Minneapolis, MN 55401 USA

Voyageur Press titles are also available at discounts in bulk quantity for industrial or sales-promotional use. For details write to Special Sales Manager at MBI Publishing Company, 400 1st Avenue North, Suite 300, Minneapolis, MN 55401 USA.

To find out more about our books, join us online at www.voyageurpress.com.

ISBN-13: 978-0-7603-2473-8

On the front cover: Photograph © Lynn M. Stone
On the back cover: top: Courtesy of William A. Suys, Jr.

Editor: Danielle J. Ibister
Designer: Maria Friedrich

Printed in China

Page 1: Single Comb White Leghorn. (Photograph © Norvia Behling)

Page 2: Lakenvelder. (Photograph © Lynn M. Stone)

Page 3: Silver Penciled Wyandottes. (Photograph © Lynn M. Stone)

Page 6: White Dorking. (Photograph © Lynn M. Stone)

Facing Page: The Brahma shares its name with the Hindu god known as the "Lord of Creation"—quite an honor in the world of chickens. (Photograph © Alan and Sandy Carey)

To Doris (2001–2005). Doris was a White Bantam Leghorn who appeared on television and the radio. She was kind and beautiful and is missed by all who knew her.

Acknowledgments

I would like to thank the many chicken experts who helped me with advice, editing, and contributing information and images, including Ric Ashcraft, Jim Finger, Barry Koffler, Jean Robocker, Hans Schippers, John Skinner, Loyl Stromberg, and a special thanks to Don Schrider from the ALBC for his editing and advice. I would also like to thank my editor Danielle Ibister for her thorough and diligent work. I am also thankful to the wonderful artists and photographers past and present whose beautiful works are featured in this book: artists Edwin Megargee, Mashime Murayama, Arthur O. Schilling, Franklane Sewell, William Suys, Diane Jacky, and photographers Thomas A. Naegele, D.O., and Martin Hintz. I am also grateful to the Steenbock Memorial Library in Madison, Wisconsin, for the use of their wonderful collection of chicken books and to the many chicken sites throughout the Internet for their input.

10
WESTERN
STONEWARE CO.
MONMOUTH ILL.

Contents

Acknowledgments 5

Introduction 8

Chapter 1
The Global Chicken 12

Chapter 2
Physical Characteristics 26

Chapter 3
Behavior 36

Chapter 4
Eggs and Chicks 44

Chapter 5
Everything But the Cluck 48

Glossary 52

Breed Classification Table 58

How to Use the Breed Profiles 60

Breed Profiles 66
Ameraucana 66
Ancona 67
Andalusian 68
Araucana 69
Aseel 70
Australorp 71
Barnevelder 72
Belgian Bearded d'Anvers 73
Belgian Bearded d'Uccle 74
Booted Bantam 75
Brahma 76
Buckeye 78
Campine 79
Catalana 80
Chantecler 81
Cochin 82
Cornish 85
Crevecoeur 86
Cubalaya 87
Delaware 88
Dominique 89
Dorking 90
Dutch Bantam 92
Faverolles 93
Frizzle 94
Hamburg 95
Holland 97
Houdan 98
Japanese Bantam 99
Java 100
Jersey Giant 101
La Flèche 102
Lakenvelder 103
Lamona 104
Langshan 105
Leghorn 106
Malay 107
Minorca 108
Modern Game 110
Naked Neck 112
New Hampshire 113
Old English Game 114
Orpington 115
Phoenix 116
Plymouth Rock 117
Polish 118
Redcap 120
Rhode Island Red 121
Rhode Island White 122
Rose Comb Bantam 123
Sebright 124
Shamo 125
Sicilian Buttercup 126
Silkie 127
Spanish 128
Sultan 129
Sumatra 130
Sussex 131
Welsummer 133
Wyandotte 134
Yokohama 137

Appendix 138
Resources 138
Bibliography 141
Index 143

Introduction

Although there are more chickens than people in the world, chicken watching is not a common activity. How often do you grab your binoculars off the windowsill to identify a chicken that has crossed the road and roamed into your backyard? Even dedicated bird watchers probably don't run across stray roosters or footloose hens wandering the windswept wilds. Even more unlikely is the possibility of encountering a flock of domestic fowl while hiking or driving through the countryside—unless, of course, you detour through a barnyard.

For the most part, chickens are housed and protected from their many predators, yet they do range free throughout the world and feral chickens do exist.

Keep this guide handy, because you never know when you'll have an auspicious sighting. You might want this guide if you're attending a state fair or poultry show to check out the wonderful array of exotic and familiar chicken breeds. You may want it in your glove compartment if you're visiting Key West or other chicken-friendly towns.

This guide is for the farmer and breeder, as well as the non-farmer who simply loves chickens. There are a myriad of reasons to be a chicken enthusiast—as many reasons as there are to use this guide. Depending on where you live and local ordinances, you may even want to start raising chickens. They are the ideal pet, needing only food, water, and a safe place to be locked up at night. Luckily, I live in a suburb of Milwaukee, Wisconsin, and our neighbors do not object to chickens. In fact, I have been raising them for twenty years!

Chicken watching can be a wonderfully satisfying, yet admittedly time-consuming, passion. Like people who are mesmerized by the graceful movements of fish within an aquarium, we chicken people are entertained for hours by the constant activities of our fair flock. Unless they are sleeping or laying eggs, chickens rarely stay still. They are constantly moving and doing chickeny things—scratching, pecking at food (or each other), taking dust baths, perching, preening, and having chicken sex. Regarding the latter, roosters won't disappoint any chicken watcher.

Facing Page: The Delaware is a dual-purpose fowl, famed both for laying large brown eggs and making tasty broilers. (Photograph © Lynn M. Stone)

Why Watch Chickens?

- It's relaxing.
- You can be out in the fresh air and get a "back to nature" feeling.
- It gives you a sense of timelessness. Most likely, some of your forefathers or foremothers spent times in their yards with their chickens.
- It's entertaining. Chickens have personalities and constantly interact with each other. They form what appear to be friendships, yet they also can be prone to bickering amongst themselves. Yes, the pecking order is alive and well.

Tips on Interacting with Chickens

- Approach chickens slowly. Chickens startle easily and any sudden movement will make them flee.
- Lure them with food. Chickens can easily be bribed with almost any food. They are great recyclers, and ours enjoy many of our leftovers.
- You can try attracting them with clucking noises, but they prefer a handful of munchables.
- Even though chickens may approach you out of curiosity, don't be discouraged if they run when you try to pet them. They have a natural tendency to flee. But with patience, you can usually train a chicken to eat out of your hand.

Chicken Details

This diagram shows the various components of the chicken's anatomy. Many of these terms are useful in identifying chicken breeds and their physical characteristics.

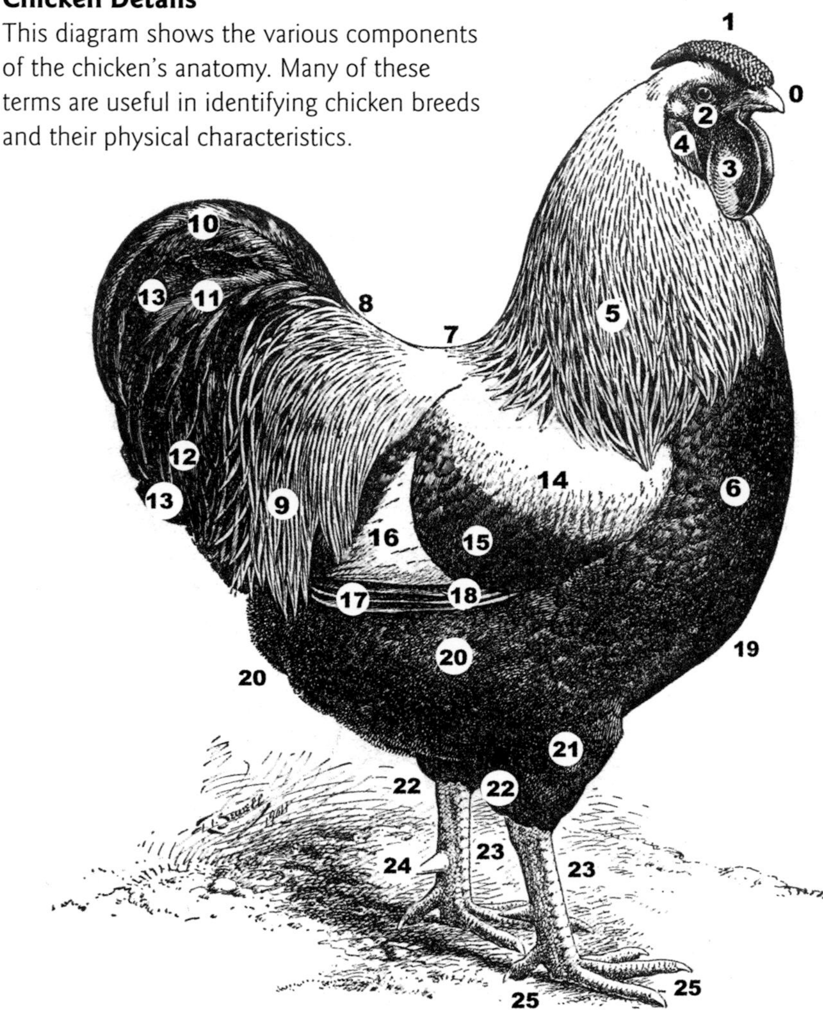

0. Beak
1. Comb
2. Face
3. Wattles
4. Earlobe
5. Hackle
6. Breast
7. Back
8. Saddle
9. Saddle feathers
10. Sickles
11. Lesser sickles
12. Tail coverts
13. Main tail feather
14. Wing-bow
15. Wing coverts
16. Secondaries, wing-bay
17. Primaries, flight feathers
18. Flight coverts
19. Keel
20. Fluff
21. Thigh
22. Knee-joint
23. Shank
24. Spur
25. Toes or claws

THE POULTRY OF THE WORL
Portraits of all known valuable breeds of fowls

Chapter 1

The Global Chicken

Around the world, there are an indeterminate number of chicken breeds and varieties. The ubiquitous chicken is raised in almost every corner of the earth except Antarctica. Describing every breed in one volume is all but impossible, so this book profiles only the breeds listed in the *American Standard of Perfection*. But apart from the Breed Profiles, it traces the migration of chickens from their origins to their arrival in America and discusses some interesting representatives from various countries.

Chickens have been flocking around since the dawn of history. It is generally agreed that peasants domesticated the red jungle fowl (*Gallus gallus*) in Thailand and Indonesia more than eight thousand years ago. Other types of jungle fowl inhabited Southeast Asia, including the gray or Sonnerat's jungle fowl (*Gallus sonnerati*) in India, the

Above: The red jungle fowl adapted to domestication over eight millenia ago and is believed to be the precursor of all of today's chicken breeds. This depiction of the red jungle fowl was created by C. H. Weigall for the *Illustrated Book of Domestic Fowl* (1884), edited by Martin Doyle.

Facing Page: This 1868 chromolithograph, first published in Boston, depicts thirty chicken breeds from around the world, including Indian Fowl, Columbian Fowl, Guinea Fowl, and Duck-winged Game.

Ceylon jungle fowl (*Gallus lafayettei*) in present-day Sri Lanka, and the green or fork-tailed jungle fowl (*Gallus varius*) in the Indonesian island of Java. But the ability of the red jungle fowl to be domesticated set it apart from the others. The bird, a small pheasant-type chicken, sported a comb and wattles and crowed like our modern domestic roosters. The birds were not used for food but for the sport of cockfighting. Fed and protected, the fowl multiplied quickly and spread throughout Southeast Asia.

Today, more than forty varieties of Thai game fowl are included in Thailand's standard of perfection. In addition, Indonesia is home to many breeds of the Ayam, *ayam* being the Indonesian word for chicken. The Ayam Bekisar is the national bird in western Java; they are kept aboard fishing boats in bamboo cages, and their loud, long crows are used by fishermen to keep in contact with other boats. The prized Ayam Cemani has black plumage highlighted by a greenish shine, black shanks and black toenails, a black beak and black tongue, black skin, a black comb and black wattles, and even black organs. In Thailand, it is believed to have mystical powers. The Kedu Cemani, another all-black chicken, is a variation of the Ayam Cemani.

Southeast Asia is also home to the Serama, the world's smallest and lightest bantam chicken. Originally from Malaysia, this bird stands only 6 to 10 inches high. The cock weighs 8 to 12 ounces, the hen 6 to 10 ounces. The cock's crow is one-third the volume of

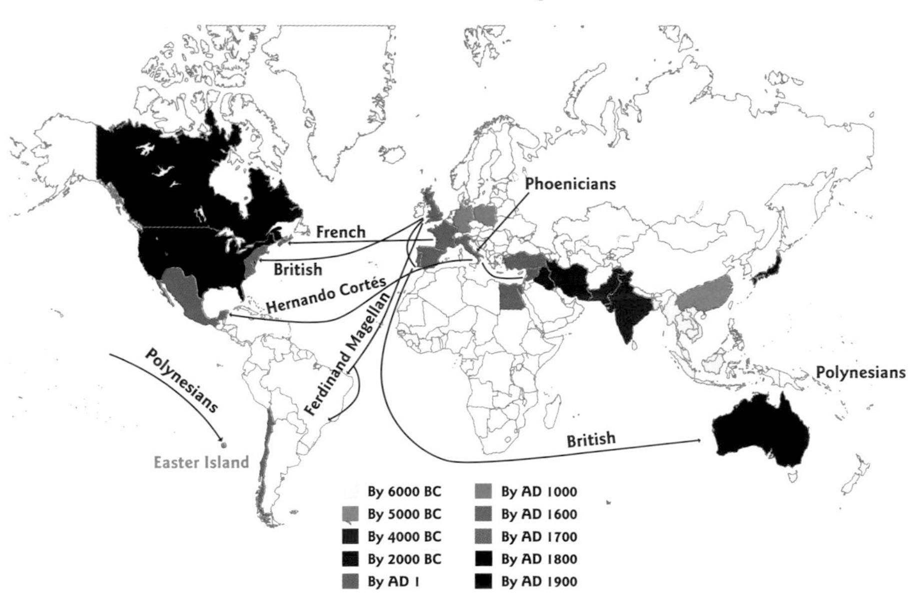

This map tracks the history of the movement of chickens throughout with world, beginning in Southeast Asia about eight thousand years ago. Today, there are chickens everywhere except Antarctica.

a regular rooster. This bird comes in all colors, and there are two thousand documented variations of hue. They are said to be wonderful pets and, bred to be people-friendly, even crave attention. Like most bantams, they are economical as well, eating only about a pound of food per month. The Serama is Malaysia's most popular household pet and has gained tremendous popularity in the United States.

By 5000 BC, domesticated chickens had spread to China from Thailand and became an important part of the Chinese culture and culinary world. In Chinese, the word for rooster (*ji*) has the same pronunciation as the word for "good luck." The rooster is the symbol of yang, one of the two life forces in Chinese philosophy. It also serves as one of the twelve animals of the Chinese zodiac cycle. According to myth, Buddha invited all the animals of the kingdom for a meeting, but only twelve showed up and each was rewarded with a year named in its honor. Emperor Huang Ti introduced the first cycle of this zodiac in 2637 BC. The first rooster year in the twenty-first century is 2005. People born in the year of the rooster are said to be lucky, highly intelligent, diligent, talkative, and sometimes prone to being "cocky."

Although chicken became part of the Chinese diet soon after the bird's arrival, it did not become a standard of Chinese cuisine until the Shang Era (1520–1030 BC). In the fifth century BC, the Chinese philosopher Confucius elevated the concept of cuisine by teaching that cooking depends on harmony of color, texture, taste, aroma, and presentation. One of his most famous dishes, "The Eight Immortals Crossing the Sea Gamboling around the Arhat," featured chicken as the first ingredient.

From China, chickens were introduced to Japan more than two thousand years ago. The Japanese excelled at developing breeds for exhibition, pets, cockfighting, and egg/meat production. Today, there are approximately fifty Japanese breeds, most of which descend from the Jidori (Japanese for "indigenous chicken"), the Shokoku, and the Shamo.

In Japan, chicken breeds became forms of living art. Breeders developed roosters with extremely long tails. The tail feathers do not molt, resulting in

Onagadoris demand high roosts to keep their beautiful tails clear of the chicken yard.

their great lengths. The famed Onagadori has a record tail length of forty feet These gorgeous breeds also include the Jitokko, Kuro Kashiwa, Minohiki, Ohiki, Phoenix, Satsumadori, Shokoku, Tomaru, and Totenko.

Chickens are held in high esteem in Japan, and seventeen Japanese chicken breeds have been designated as national treasures. These include, among others, the long-tailed Onagadori (Japanese for "honorable fowl"); the twenty varieties of Chabo or Japanese Bantam; the Ko-Shamo (*Ko* meaning "miniature"); and the Kuro Gashiwa, characterized by dense black feathers. The Oh-Shamo was introduced from Thailand in the early seventeenth century for cockfighting; today, it is raised commercially for its finely textured meat. Also originally bred for cockfighting, the Shokoku and Satsumadori are now raised for their beauty. Other Japanese national treasures include the Minohiki, characterized by a full and long tail, the Koeyoshi and the Tomaru (both long crowers), and the silky-feathered Ukokkei.

From the Far East, chickens gradually spread north, west, and east, and by 4000 BC, they were in Harappa, a farming region along the Indus River in Pakistan. Their presence is evidenced by two ancient clay figurines, one of a cock and one of a hen, from that period.

The fowl also made history in ancient Egypt, and evidence indicates that they existed there well before Christ. Archaeologists found a chicken icon in King Tutankhamen's tomb dating from circa 1400 BC. In the fourth century BC, Egyptians started to mass-produce chickens in large "hatching ovens." These structures had thick, mud-brick walls and varied in size; some were large enough to hatch fifteen thousand eggs at one time. The ovens were housed in buildings that also had nesting areas for hens, pens for the newborn chicks, and quarters for the operator. Sun-dried camel dung and straw were used for fuel. The operator worked without a thermometer, using intuition to keep the hatching ovens at the proper level.

The Fayoumi, or Bigawine, one of the ancient Egyptian breeds, gained fame for its egg production. The popular bird of Egypt and Palestine during the time of Jesus, it may have been the breed of cock that crowed after the third time the disciple Peter denied Christ. Although their feathering is penciled, the chicks hatch brown before turning silver or golden. Today, Fayoumis fend for themselves in Egypt and continue to lay eggs at a high rate.

Domesticated fowl migrated through Persia (now Iran). Today, Iran maintains a wide variety of chicken breeds with such exotic names as the Chity, Hanayee, Kabotary, Siyahe Kantony, Talayee, Zireh E., and Manx Rumpies. When the chicken arrived in Greece, the Greeks referred to it as the "Persian bird" and made it the subject of myth, science, art, and literature. As early as the sixth century BC, fabulist Aesop, a Greek slave, used stories of the rooster and the hen to impart his wisdom. The great philosophers Socrates, Plato, and Aristotle,

Though the Russian Orloff originated in northern Iran, it was developed in Russia and named after Count Orloff Techesmensky. In the 1800s, the birds made their way to Europe, where they were perfected by breeders in Germany and popularized in Holland and Great Britain. Today, the ABA and the British Standards recognize certain varieties. The Orloff bears a small walnut comb, perhaps developed to avoid frostbite in the harsh Russian winters.

living in the fifth and fourth centuries BC, also commented on the chicken.

Chickens spread by land and water throughout Europe. As early as 500 BC, Phoenician traders hauled cargos of the precious fowl on their Mediterranean voyages. The birds also played an important role in ancient Etruria, the country that inhabited present-day Italy prior to its defeat by the Romans in the sixth century BC. The Etruscans used chickens to foretell the future by a divination method called alectromancy, where priests interpreted the future by the way a chicken ate corn. Under Roman rule, chickens became integral to daily life, continuing to be used in religious rituals. The Romans also made great progress in poultry science and gastronomy.

During the height of the Roman Empire, the chicken was considered sacred and valued for its supposed medicinal qualities. Chickens were studied, scrutinized, and written about by great scholars such as Columella (first century AD), Varro (116–29 BC), and Pliny the Elder (AD 23–79). Later, during the Italian Renaissance, the naturalist Ulisse Aldrovandi (AD 1522–1605) wrote extensively about the entire body of chicken wisdom in his book *Historia Animalium*.

When Julius Caesar invaded Great Britain in 55 BC, his troops discovered chickens already established on the island. The birds had probably been brought to the distant outpost by hardy Phoenician traders. At the time, the chickens were used solely for cockfighting. The Romans also found the birds well ensconced in the areas of France and Germany.

On its journeys to the West, the chicken has crossed many roads. It arrived in the Netherlands, Belgium, and Germany as early as the Middle Ages (AD 476–1453), brought to these countries by Dutch and other seafarers. In the Netherlands, most farmers in and around Barneveld kept poultry as early as the twelfth and thirteenth centuries. In the mid-1300s, the Duke of Brabant sent the bailiff from town to town to collect land rents, which tenants often paid in chickens. Most famous for the development of the Penciled Hamburg breed, the Netherlands have a rich poultry history. The Nederlandse Hoender Club, a special breed organization in the Netherlands, has been actively promoting the Dutch national breeds for more than a hundred years. Today, there are

Aldrovandi's Chickens in 1600
Left to right:
Top: Indian rooster, Swamp rooster, Male Paduan
Middle: Persian rooster, Turkish hen, Bantam hen
Bottom: Wool-bearing hen, Rooster with feathered feet

more than three hundred Dutch breeds and other varieties.

Belgium also has a long history of breeding chickens, including the Braekel, Campine, and Malines. The oldest is the Braekel, whose ancestors existed in Flanders as early as the twelfth century. A large bird, it was bred for both egg production and meat. A society formed in Nederbrakel in 1898. As with many European breeds, the Braekel population was devastated by wars and neared extinction. The breed revived with the help of a club formed in 1971. The Campine is similar to the Braekel though smaller. A Braekel-Campine fowl was brought to the United States and admitted to the *American Standard* in 1894 but dropped by 1898. The Campine, admitted to the *American Standard* in 1914, was developed primarily for egg production. The Malines, also called Poulardes de Bruxelles and Mechels, was bred in an area of Belgium called Mechelen. Two varieties that exist today are the Cuckoo, which has a pea comb, and the Turkey-Headed, known for its unique triple comb.

The Turkey-Headed Malines is a native of Belgium.

The British, who documented chickens as early as the 1500s, helped spread them to other parts of the world. In 1787, a fleet of eleven ships carrying 759 convicts set sail from Portsmouth, England, bound for Australia. They brought chickens, as well as seeds, cattle, sheep, pigs, goats, horses, and two years' supply of food to help establish the new settlement. Today, the most popular breeds in Australia are the Australorp; the Australian Langshan, a smaller version of the Croad Langshan; and the Australian Gamefowl, once called Colonial Game. They were developed there in the late nineteenth and early twentieth centuries as fighting fowl for the colonial soldiers on duty guarding convicts.

When the British Parliament banned cockfighting in 1849, poultry exhibitions began. A poultry club formed in 1863, and the first book of poultry standards was printed in 1865. The Poultry Club of Great Britain, dating back to 1877, oversees the British Poultry Standards set forth by the specialist breed clubs. The British Poultry Standards includes eighty-six varieties of large fowl and bantams, about twenty-five more than are accepted in the *American Standard of Perfection*.

The French also have a longstanding love affair with the *coq*. In fact, the Latin term for chicken, *Gallus*, means both "France" and "cock." The Romans

The Dutch Breeds

The Hamburg is one of Holland's most enduring chicken breeds. The *American Standard* recognizes six varieties, including the Silver Penciled, pictured here.

The Penciled Hamburg descends from the Friesland Fowl, one of Holland's oldest breeds. Prolific layers of white eggs, Friesland hens were referred to as "everyday layers." Though the Friesland usually had a single comb, there was also a rose-comb variety of the Silver and Golden Friesland. These were called Hollands Hoens (Dutch Fowl), or Hollanders.

The Chaam Fowl, or Chaams Hoen, was a common dual-purpose breed that neared extinction and was revived around 1900. There are two varieties, Silver Penciled and Golden Penciled, having black bands equal in width to the silver and gold. They have unique orange eyes and slate blue legs. (Courtesy of Hans Schippers)

The Brabanter is portrayed in many paintings dating from the seventeenth and eighteenth centuries. The oldest known is the 1676 work by Melchior d'Hondecoeter, famous for his beautiful poultry paintings. The most important characteristic of this breed is its "helmet crest," which stands up but is flattened sideways and toward the front of the head. (Courtesy of Hans Schippers)

A very old breed, the Drenthe was given its name around 1900 by the famous Dutch poultry specialist R. Houwink Hzn. The breed was from the province of Drenthe. They are unusual in that they are rumpless, except for the Partridge variety. (Courtesy of Hans Schippers)

The Dutch Owlbeard is one of the oldest Dutch breeds. It is known for its unique comb with two spikes. The fowl has no wattles but sports a long, full, rounded beard. The most striking variety is the Moorkop, or Moorhead, which has body feathers in white, buff, or blue, and a black head and beard. (Courtesy of Hans Schippers)

The Breda Fowl descends from different crested breeds. A feather-footed variety is featured in the 1660 Dutch painting *De Hoenderhof*, or *The Chicken Run*. The first reliable written descriptions are from the mid-nineteenth century. The Breda, or Kraaikop (translated "Crowhead"), got its name because its head looks like a crow's, even lacking a rudimentary comb. (Courtesy of Hans Schippers)

gave the name to France because of the country's abundance of chickens. The *coq Gaulois* ("Gallic rooster") is the national symbol of France, and the *poulet* ("chicken") the pride of the French gourmands. The first chicken breeds established in France were the Gauloise Doree, the Ardennaise, the Alsacienne, and the Combattant du Nord. With the arrival of the Asiatic breeds in the last half of the nineteenth century, a frenzy of development occurred, creating the Houdan, Faverolles, Crevecouer, and others. The French established a standard in the early 1900s and, today, recognize more than sixty breeds.

Another popular breed in France is the Marans, famous for its chocolate-colored eggs. The bird hails from the town of Marans, a busy port north of La Rochelle on the Atlantic coast. As early as the twelfth century, British ships arrived there and traded supplies, including chickens. The Marans was originally a cross of gamecocks from England and hens from Marans. By 1880, the popularity of the dark brown eggs soared and the English started breeding Marans, but the quality became compromised. Then, around 1920, Madam Rousseau, near Marans, began to control the quality of Marans eggs, and in 1929, the breed was presented at the Poultry Breeders Society in Paris.

Chickens traveled east with Polynesian seafarers as they crossed the

First imported from Shanghai, China, in the 1840s, Cochins helped launch the poultry exhibition movement in Great Britain and the United States. Queen Victoria herself exhibited prize Cochins in the royal poultry yard, as depicted in this illustration titled "Her Majesty's Cochins." The early Cochins did not have the round appearance of today's Cochins, created by a profusion of plumage and down fiber in the under fluff, nor did they have the feathered legs of modern Cochins.

Pacific. The birds arrived in the Easter Islands and were domesticated there in the twelfth century. Ferdinand Magellan found chickens when he arrived in South America in 1519, as did Spanish conquistador Hernando Cortés when he invaded Mexico in 1520. Chickens spread north to the present-day United States in the mid-1500s, brought by the conquistadors when they invaded the area of New Mexico. Chickens arrived in Canada with the French in the 1700s when they settled in Louisbourg, Nova Scotia.

Chickens also made the arduous Atlantic voyage to Jamestown, Virginia, arriving with the colonists in 1607 and later with the Pilgrims. These birds were a motley bunch and primarily used as egg layers and gamecocks. Chickens, at this point, were considered inferior to other livestock. Until the 1830s, poultry breeding was done on a more or less accidental basis, although farmers did improve birds' shape and color by selecting roosters from the best flocks to mate with their hens. Certain areas gradually became known for their prized poultry. Breeders in Rhode Island and southern Massachusetts raised large reddish chickens, the ancestors of today's Rhode Island Reds. After World War II, the chicken industry became big business, using crossbreeding to improve egg-laying ability and meat quality.

In the 1840s, the American poultry

This Currier & Ives lithograph captures the spirit of the gamecock. Cockfighting inspired the domestication of the red jungle fowl thousands of years ago. Banned from fighting in most countries by the end of the nineteenth century, the gamecock is now a key player in the world of poultry exhibition. (Courtesy of the Library of Congress)

Postal Chickens

Chickens are honored on stamps worldwide and collected by many chicken philatelists. (Courtesy of Norman V. Saari)

world was revolutionized by the importation of new and exciting Asiatic breeds: the Shanghai and the Chittagong, also known as the Cochin and the Brahma. Even the professional classes—doctors, lawyers, politicians—began to take an interest in breeding poultry. "Hen fever" swept the nation, as described in chicken breeder George Burnham's *The History of the Hen Fever: A Humorous Record* (1855). When Boston held the first "Exhibition of Fancy Poultry in the United States of America" in 1849, the American poultry industry was truly launched. Ten thousand people came to witness the exotic fowl of 219 exhibitors. These breeders took the poultry world by storm. New breeds were soon being created and existing breeds perfected for their beauty, size, meat quality, and egg-laying ability.

In 1873, the first American Poultry Association formed in Buffalo, New York, to organize the myriad poultry fanciers. A year later, the association adopted the first *American Standard of Excellence* in order "to standardize the varieties of domestic fowl so that a fair decision could be made as to which qualities marked prize winners." This guide described forty-one large fowl varieties and twenty bantam varieties prevalent in the United States. In 1905, the guide changed to the *American Standard of Perfection*. Now, the *American Standard of Perfection* details fifty-three breeds of large fowl, sixty-one breeds of bantams, and hundreds of varieties.

The American Bantam Association (ABA) is another important poultry organization. Founded in 1914, the organization was formed to represent the growing number of bantam breeders. The ABA *Bantam Standard* gives full descriptions of fifty-seven breeds, eighty-five plumage patterns, and hundreds of varieties of bantams.

Today, chickens enjoy a growing popularity. An Internet search yields more than four million chicken-related sites, covering topics from chicken clubs to chicken shows. More and more chicken books, such as *The Field Guide to Chickens*, are published annually. Chicken collectibles are in high demand. Individual chicken breeders and large companies sell chicks that can be shipped to customers overnight. More urban areas are becoming chicken friendly, such as Portland, Oregon, which allows three hens per household. As the chicken world continues expanding, this field guide will be a helpful addition to your library.

Chapter 2

Physical Characteristics

Before we discuss the intricacies of chicken identification, certain generalities need clarification. Though different breeds vary in size, shape, and color, they have some commonalities. Generally speaking, all chickens have relatively small heads with strong beaks, bodies covered with feathers, and breasts shaped like the keel of a boat. They have tails (with the exception of the Araucana and some other rumpless breeds), wattles, a visible comb of some sort (except the Polish, which has a V-shaped comb hidden in its crest), two strong scaly legs, usually four toes (and occasionally five) with sharp claws, and short wings. The male is larger than his female counterpart and often more flamboyant in his coloration. Chickens peck, but they can't bite because they don't have teeth. Surprisingly, the male and female external sex organs—or in the cock's case, the lack thereof—are similar. Each has a cloaca, or hole, through which the sperm is transmitted from the rooster and deposited in the hen.

Combs, feathers, and colors are the three major identifying features of chickens and are discussed in detail in this chapter. After combs, wattles are the second major feature that differentiates a chicken from most other birds. These are the two fleshy appendages that dangle under the chin. Like the comb, the male's wattles are larger than the female's.

Like the comb and wattles, the male's tail is larger than the female's. The tails vary from dragging on the ground to standing totally erect, from being small (Cochin and Brahma) to having a tail up to forty feet long (the Japanese Onagadori). The angle of the tail in relation to the horizontal helps in their identification.

Facing Page: Twisted feathers characterize the Frizzle, a variety that exists in several breeds. (Photograph © Norvia Behling)

Combs

One of the features that differentiates a chicken from other birds is its crowning glory—the comb. This reddish, fleshy appendage sits on top of the head, and the male's is more prominent than the female's. Combs come in different sizes and shapes and act as one of nature's methods to attract the opposite sex. Most chickens have combs, with the exception of some exhibition breeds and game fowl that are dubbed, or trimmed, for show. Combs are typically red, with a few exceptions. Sumatras, Birchen, Brown Red Modern Games, and Silkies have purple combs, and Sebrights have purplish red ones. Combs are soft to the touch and can blacken from frostbite in extremely cold weather.

The shape of combs varies according to the breeds. The following are the different types.

Buttercup Comb

buttercup: Cup-shaped comb that starts at the top of the beak with a single ridge and then forms a circle of even points that looks like antlers. The buttercup comb looks like a glorious crown atop the chicken's head and is typical of the Sicilian Buttercup breed.

carnation or **king's**: Unusual single comb with several lobes at the rear. It is typical of the Penedesenca, a Spanish breed. In Spain, this comb is called *cresta en clavel* ("carnation comb") or *cresta de rei* ("king's comb").

cushion: Small, smooth, round, knob-like comb planted on top of the beak and extending partway up the head. The cushion comb is typical of the Chantecler.

Cushion Comb

Pea Comb

pea: Medium-sized comb with three ridges running lengthwise from the top of the beak to the top of the head. Though rather low, the middle ridge is larger with slight serrations or undulations. The pea comb is typical of the Araucana and Ameraucana, Aseel, Brahma, Buckeye, Cornish, Cubalaya, Sumatra, and Shamo.

Rose Comb

rose: Tube-shaped comb with small rounded bumps extending from the top of the beak to the back of the head. It ends in a pointed spike. When viewed from above, it looks like a teardrop. The rose comb is typical of Dominiques, Hamburgs, Rose Comb Leghorns, Wyandottes, Rose Comb Black and White Minorcas, Red Caps, Rose Comb Bantams, Sebrights, Rose Comb Rhode Island Reds, and Rhode Island Whites. In many breeds, like Wyandottes, the spike remains close to the head.

single: Most common of the comb styles, being a thin, single line starting at the top of the beak and extending to the top of the skull. It sports five to six points between deep serrations, the middle point being the largest. The male's is always erect, while the female's lops over in some breeds.

Single Comb

Single Lopped Comb

strawberry: Cushion comb with bumps. The strawberry comb is typical of Malays and Yokohamas.

Strawberry Comb

V-shaped: Comb consisting of two horns joined at the base. The V-shaped comb is found on the Houdan, Polish, Crevecoeurs, La Flèche, and Sultans. It is also called the antler comb or the horn comb.

V-shaped Comb

walnut: Almost round comb but a little wider than it is long. It is somewhat lumpy and has a narrow transverse indentation slightly to the front. The walnut comb is typical of the Silkie, although the bird's large crest sometimes hides the comb.

Feathers

Feathers are appendages growing out of the chicken's skin and forming plumage. They protect the chicken from the elements—rain, cold, sun. Feathers are composed of webbing, which is barbs that stick together and form a smooth sheet; the quill, also called the shaft or barrel, which is the hollow base of the feather; and the down, which consists of barbs that are not united.

Feather patterns are important in identifying chickens.

Barred feathers have horizontal stripes of two different colors. These bars may be regular or irregular, depending on the breed. Barred Plymouth Rocks have feathers that are regular, having bars of approximately the same width. Irregular barring, also referred to as cuckoo, has one bar of color more dominant than the other and is typical of Dominiques, Hollands, Campines, and others.

Cuckoo feathers have irregular dark and light bars with the tips of each feather dark, as in Belgian Bearded d'Anvers and Dorkings.

Laced feathers have a narrow border of a contrasting color around the web of the feather. Lacing is found in nearly all Blue varieties, as well as certain varieties of Dominiques, Cochins, Cornish, Polish, Wyandottes, and Sebrights.

The **Mille Fleur** pattern has a mahogany ground color. Each feather is marked with a crescent-shaped black bar and tipped with a V-shaped white spangle, as in Booted Bantams, Belgian Bearded d'Uccles, and Belgian Bearded d'Anvers.

Mottled feathers have white tips at the ends, although the tips do not appear on every feather. This pattern is found in Anconas, Houdans, Javas, Japanese Bantams and Belgian Bearded d'Anvers.

Penciled feathers have crosswise bars. They are typical of certain female feathers, including those on the Penciled Hamburgs. Other pencilings, instead of being horizontal, are stripes that follow the edge of the feather, as in the Silver Penciled and Partridge varieties (Plymouth Rocks, Wyandottes, Cochins, Hamburgs), and Dark Cornish. To meet the *Standard* requirements, each feather in the back, breast, body, wing bows, and thighs should have three or more pencilings.

Peppered feathers have small gray or black dots. They are a defect in the *American Standard*.

Stippled feathers have dots in a color that contrasts the background color. Brown Leghorn females have stippling.

Spangled feathers are tipped with a black and/or white teardrop-shaped marking. The Speckled Sussex has feathers tipped with a white spangle. Hamburgs, Old English Games, Malays, and Aseels have spangled varieties.

Splashed feathers have irregular splashes of contrasting color and occur in mottled varieties.

Striped feathers have a solid area that runs along the center and is surrounded by lacing. The male hackles of the Silver Penciled Wyandottes have striped feathering, as do the saddle feathers on the Dark Brown Leghorn male.

Ticking refers to the cluster of small black dots on the tips of lower neck feathers in New Hampshire females.

Twisted feathers are those in which the shaft and web are twisted, creating the frizzled feathers characteristic of the Frizzle breed.

Laced Feathers

1. Buff Laced Polish
2. White Laced Red Cornish
3. Silver Laced Wyandotte
4. Light Brahmas
5. Golden Laced Wyandotte
6. Blue Andalusian
7. Golden Sebright

Penciled Feathers

8. Dark Cornish
9. Partridge Cochin
10. Silver Penciled Plymouth Rock

Spangled Feathers

11. Ancona
12. Mille Fleur Bantam
13. Silver Spangled Hamburg
14. Golden Spangled Hamburg

Barred Feathers

15. Barred Plymouth Rock
16. Buttercup
17. Silver Campine
18. Dominique
19. Golden Penciled Hamburg
20. Silver Penciled Hamburg
21. Barred Plymouth Rock pullet

Colors

Color-wise, a chicken is a thing of beauty. But the variety of colors and the intricate feather patterns have not happened by chance. Judicious and patient breeding by chicken fanciers has resulted in the wide range of colors. The basic colors of poultry plumage are black, white, blue, buff, red, and various shades of yellow. From these, a plethora of colors and patterns emerge. Some of the colors and patterns that typify certain varieties are as follows.

bay: Golden brown.

birchen: Black and pure white coloration. The hackles of the male and female and the saddles of the male have silvery white feathers with a narrow black stripe through the middle of each feather. The Modern Game and the Cochin Bantam have a Birchen variety.

black: Lustrous greenish black.

black-breasted red: Black everywhere except for red hackles, back, shoulder, and part of the wing, as in the coloring of the red jungle fowl. The male has a glossy black breast, while the female has a salmon breast with yellow-orange hackles and brown body feathers. Light Brown Leghorns, Modern Games, Dutch Bantams, and Old English Games are all technically Black-Breasted Red in color. Though the Aseel, Cubalaya, Malay, and Sumatra have a black-breasted variety, they are actually wheaten in color. The Wheaten male is almost identical to the Black-Breasted Red male, though usually with more of an even orange color in the neck and saddle feathers.

blue: The *American Standard* describes *blue* as "a term loosely used in referring to the general slaty color of some varieties of chickens." The blue varieties—such as in Ameraucanas, Andalusians, Blue Belgian Bearded d'Anvers, Cochins, Langshans, Orpingtons, and Plymouth Rocks—have feathers that are laced with glossy black. Various other blue shades are leaden blue, lemon blue (Old English and Modern Games), and bluish slate. The shade called self blue in the United States is called lavender in Britain.

The Black Sumatra is one of nearly fifty black varieties in the *Standard*. (Photograph © Lynn M. Stone)

brown: There are no longer any Brown varieties in the *American Standard*. What were once called Brown Leghorns branched in two directions in the late 1800s and now come in Light Brown and Dark Brown varieties. This color was named after a Mr. Brown who was an early American breeder of Brown Leghorns in New England. Now, the Light Brown color, found in Leghorn and Dutch Bantams, is closely related to Black-Breasted Red. The Dark Brown Leghorn is similar to the Partridge, but with stippled feathers on the female instead of laced. Brown Cochins should actually be called Brown Red Cochins.

brown red: Birchen with an orange feather instead of silvery white, as in the Ameraucana, Japanese Bantam, and Modern and Old Game breeds.

buff: Even shade of orange-yellow with a rich golden hue. The buff varieties differ. Some are completely buff, such as Buff Ameraucanas, Cochins, Cornish, Leghorns, Orpingtons, Minorcas (single comb), Naked Necks, Plymouth Rocks, Silkies, and Wyandottes. Others—such as Buff Brahmas, Japanese Bantams, and Catalanas—have some black accents. In the Black-Tailed pattern, as in Black-Tailed Buff Japanese Bantams, the tail, neck, and wings have some black.

chestnut: Dark red-brown, darker than bay.

cinnamon: Dark reddish buff.

This 1911 illustration depicts Brown Leghorns. Today, Leghorns have been more finely bred into Light Brown and Dark Brown varieties.

Columbian: A color pattern dominated by white and black. The head is silvery white. The front of the neck is white and the hackles are a lustrous greenish black with narrow silvery white lacing with a larger black shaft. The back and thighs are white. The male's main tail is black with the sickles a lustrous greenish black and coverts laced with silvery white. The female has a white body with a neck and some tail feathers black with white lacing. The Plymouth Rock, Sussex, and Wyandotte have a Columbian variety. The Light Brahma technically fits the Columbian coloration.

crele: A blend of cuckoo (indistinct barring) and Black-Breasted Red, the result being a cuckoo-feathered rooster that has some yellow, orange, and red in the hackle, shoulders, and saddles (hackle only in females). The Old English Game has a Crele variety.

duckwing: Characterized by a distinct bar across the wing of the male, as in the Araucana, Modern Game, and Old English Game.

fawn: Light brownish tan.

gipsy or **mulberry**: Dark purple, found on the face, comb, and wattles of Silkies, Sumatras, Sebrights, and other breeds. Exposure to full sunlight often intensifies this effect.

horn: Dark, brownish color that usually occurs on beaks and sometimes on legs and feet of a yellow- or white-legged breed (with a color pattern that includes black in the plumage).

mahogany: Deep reddish brown.

partridge: A color pattern imitative of partridges. The head is a lustrous rich red. The front of the neck is black. The hackles and back are a lustrous greenish black with rich red lacing. The male's tail is black, with greenish black sickles and the coverts laced with red. The female has a deep reddish head and black penciling on each feather of the back, breast, body wing bows, and thighs. Some of the Partridge varieties are the Chantecler, Cochin, Plymouth Rock, and Wyandotte.

porcelain: Straw-colored feathers tipped with various kinds of white spangles and a pale blue stripe through part of the length of the feather. Porcelain varieties exist for the Belgian Bearded d'Anvers, the Belgian Bearded d'Uccle, and the Booted Bantam.

quail: A color pattern imitative of quails. Black neck, back, and saddles feathers are laced with a golden bay. The muff and breast are a brownish yellow. The male has a black tail and the female a black tail with a light brown edge, as in the Belgian Bearded d'Anvers.

red: Rich dark red or mahogany red, as in the Rhode Island Red and Buckeye. Naked Necks and Sussex have a Red variety where the male has a glossy red neck, back, and saddle. The breast and tail are lustrous greenish black and there is a bluish black bar on the wing. Female Reds have varying tones of gold, rich salmon, browns, and black. Red Dorkings are an exception, being a variant of Black-Breasted Red.

red pyle: A color pattern of some Malays, Modern Games, and Old English Games. The male's neck is a lustrous light orange, the wing bows are a lustrous rich red, and the overall body color is white. The female has a golden head with tinges of salmon throughout.

salmon: Reddish or pinkish buff—the color of cooked salmon. It is used to describe the coloration of the breast in Black-Breasted Red females (and Light Brown Leghorns and Dutch). In Salmon varieties, the female has a white breast and fluff and the rest of the body a salmon color. The male's coloration is more diverse. Though the Faverolles has a Salmon variety, it is actually a variant on Wheaten coloration, with female more salmon and the male more golden.

self-colored: Of one even shade of color, as in Self Blue, where there is no darker blue lacing.

silver: A color pattern defined by silvery white. The male has varying shades of silvery white, black, and greenish black feathers. There is a wide greenish black bar across the wing. The female has a reddish salmon breast, a gray body, and a neck that is silvery white with a narrow black stripe down the middle of each feather. The Dorking, Phoenix, Leghorn have Silver varieties.

silver-penciled: A color pattern defined by its silvery white and penciled feathers. The male has a silvery white head. The front of the neck is black with the hackles and back a lustrous greenish black laced with silvery white. The main tail and breast is black, the back having greenish-black feathers with silvery white lacing. The female's head is silvery gray and the neck has black feathers, with steel-gray penciling and silvery white lacing. The rest of the body has an overall look of steel gray with black-penciled feathers. Plymouth Rocks and Wyandottes have Silver-Penciled varieties.

slate: Dark bluish slate, nearly black, found on the legs of certain breeds. Slate leg color is caused by a clear or dark skin layer at the surface with a black pigment below the surface. When the top layer is clear or white, the slate is blue or leaden blue. When the top layer has dark pigment, the slate is nearly black.

wheaten: Yellowish ochre, the color of wheat. In Wheaten varieties, the female is predominately wheat-colored with a darker neck, while the male's coloration is more diverse The male is similar to Black-Breasted Red but a more even shade of orange—instead of red—in the neck, back, and saddle feathers. The female's breast is cinnamon. The Aseel, Japanese Bantam, Malay, Old English Game, and Shamo have Wheaten varieties.

white: All parts of the feather in all sections are white in White varieties. The White variety is the most prevalent in the *Standard*.

willow: Yellowish green, found on the legs of certain breeds, such as Javas. Willow is caused by crossing a bird with slate legs to one with yellow legs. The lower or blackish layer of color is masked by the yellow pigment in the surface layer and causes the yellow to appear green.

Chapter 3

Behavior

When observing chickens, look at the characteristics that make them stand out from other animals. Despite a common misconception, chickens are quite smart, and they can actually be trained! I visited a chicken training class in Hot Springs, Arkansas, taught by noted animal trainer Bob Bailey. While there, I was impressed with the chicken's ability to distinguish colors and shapes, pull a string, and even dance. Naturally, however, it took a great deal of patience on the part of the chicken trainers, who used a clicker as a signal and food as a positive reinforcement.

In a more natural setting, chickens are resourceful when it comes to finding food. They know not to stray too far from the coop and NEVER cross a busy road.

They form friendships. On one occasion, one of our Rhode Island Red hens became the protector of a bantam who had been harassed by other chickens. She stayed with her by day and at night slept with her wing wrapped around her small companion. Another time, Cluck, a beautifully feathered Polish hen, was attacked by a hawk and badly injured. After she recovered, she became agoraphobic and would not leave the coop, so her hen friends took turns staying inside with her. Interestingly, in our diverse collection of chickens, friendships or cliques are often based on color or breed.

Facing Page: A White Dorking rooster communicates by crowing. (Photograph © Lynn M. Stone)

Communication

Chickens communicate effectively, and studies have determined that they make more than thirty different sounds. Most recognizable is the rooster's crow. Cockerels start practicing their crow at three to four months. Their voice starts with a squeak, like that of pubescent boys, and later develops into a proper crowing sound. The volume of the crow depends on the size of the rooster, the larger often sounding like a donkey's bray, while the smaller has a more high-pitched tone. Roosters also make a low-pitched, threatening sound to warn hens of danger and a friendlier, high-pitched sound to beckon them.

Hens are more varied in their communication. Chicks and their mothers communicate with a variety of different sounds. The chicks cheep, while their mothers cluck. If separated, the chick makes a high-pitched tone of panic, to which the mother responds with a faster clucking tone. Hens cluck gently as a sign of contentment and make a *gagaga* sound when alarmed. Before laying an egg, they produce a complaining, cawing sound and, after laying an egg, proudly announce their prize with a cackle.

Though not common in the United States, some roosters can sustain a crow for several minutes instead of the usual few seconds. Long crowers include foreign breeds such as the Berat Fowl, Bergische Kraeher, Denizli, Koeyoshi, Kuro Gashiwa, Pelung, Tomaru, Totenko, and Yurlower.

Eating

Eating is a common chicken activity. Chickens can recognize a grain of food from a distance of about three feet. When eating, the chicken hones in on the grain, getting about one and a half inches from it and then pecks at it. The chicken bobs its head because it is actually getting the grain in its vision each time before pecking. In addition to eating grains, chickens are also predators to worms, bugs, and an occasional mouse.

An Australorp focuses on a grain of food before pecking at it. (Photograph © Norvia Behling)

A Barred Plymouth Rock flaps its wings. Most large breed chickens are too heavy to fly. (Photograph © Norvia Behling)

Flying

Even though chickens have wings, they have essentially lost the ability to sustain flight. Depending on their weight, heavy chickens such as Cornish can do little more that flap their wings and barely jump. Lighter chickens, especially bantams, can fly short distances and have no trouble clearing a tall fence. When frightened, many varieties fly high into trees. Most chickens are adept at flying straight up in the air like a helicopter to reach their roost.

Hearing

Chickens have a well-developed sense of hearing, a necessity for detecting predators. There are no visible ears; their auditory canal is hidden by feathers and a flap of skin.

Hygiene

Though chickens have the reputation for being dirty, due to a frequent malodor in the chicken coop, they do have a natural instinct for cleanliness. Watching a chicken dust bathe is quite entertaining. They usually find a sunny dry spot, scratch an area of dry soil, and wriggle in the dust to rid themselves of parasites, seeming to luxuriate in the activity.

Pecking Order

The much-used phrase "pecking order" was originally coined by Norwegian psychologist T. Schjelderup-Ebbe, and chickens were its basis. The term describes a social hierarchy of dominance and subordination in a group. In the world of chickens, the pecking order starts early. From as young as two weeks, chicks peck playfully at each other. Toward puberty—usually ten to twelve weeks for pullets and twelve to sixteen for cockerels—more serious bickering ensues, until the real pecking order is established at about twenty-six weeks.

If there is one rooster in the flock, he is most often the king of the roost. The hens will vie for his attention, using force if necessary by pecking their opponents around the face. Add more roosters to the flock and trouble ensues. There is often chasing, quarreling, and fighting amongst the entire flock. Presently, we have five roosters and twenty-five hens. Two of the boys constantly vie for top position, fighting over the hens. The poor rooster on the bottom of the pecking order is bullied and the others currently will not allow him to sleep in the same coop with them. Pecking order among roosters is not necessarily dependent on size. I have had fierce bantam roosters who ruled the roost.

If there is no rooster in the group, the hens will enter into fierce battle to establish which is queen. A hen's comb size (bigger is better) can establish her as superior to the other smaller-combed rivals. This hen will take on the male role in the flock. She may even crow and try to mount the other hens.

If a cockerel is introduced into a roosterless flock, he is often intimidated by the aggressive hens and has no chance of mating with them.

Perching

Chickens sleep upright and are happiest on a perch. When sleeping, their toes grip the perch, and their muscles lock into place so they don't fall off. The usual sleeping position is with the head tucked under the wing. The instinct to perch is a throwback to chickens' wild relatives, who perched in trees.

Sex

Chickens mate throughout the year, and there is no monogamy in the world of chickens. Though chickens sometimes mate throughout the day, they most often copulate in the afternoon when there is a lower probability of an egg being in the oviduct, thus increasing the odds of fertilization. Some roosters will mate with anything that doesn't move quickly enough. Chicken sex is quite a sight. When the rooster approaches his target, he stands erect and fluffs out his neck feathers. He then dances around her with the wing closest to her spread out and lowered. This movement is sometimes called the waltz or wingflutter. Then he pounces. He jumps on the back of the hen, using his wings to balance, pecking her neck to get a grasp and treading, pushing her down with his feet. The hen lowers her head and raises her tail as the rooster presses his cloaca against hers, spewing sperm. This process is called the "Cloacal Kiss" and takes about ten seconds. Often, though, a hen is not in the mood. She escapes by running, stepping aside, crouching, or moving her tail to one side. If the act has been completed, the rooster may step off, circle and waltz again, or crow boastfully. The hen often stands up, shakes, and runs away.

A broody hen turns her eggs several times a day. (Photograph © Norvia Behling)

Broody Hens

Broody hens are wannabee chicken mothers. When a hen decides that motherhood is in the cards, her behavior changes. For the twenty-one days it takes until the eggs hatch, she sits on her nest, fiercely defending her clutch, and ruffling her neck feathers and squawking if disturbed. She leaves the nest only once a day for food, water, and her daily constitutional. A broody hen seems in a trance. She clucks away, turning her eggs two or three times a day, and waits patiently for the blessed event. Certain breeds, including Cochins, Frizzles, and Silkies, are more prone to broodiness. Hens disinclined toward broodiness are referred to as "non-sitters."

Vision

Chickens have eyes on either side of the head. To see things in three dimensions, chickens need to focus first with their left eye and then with the right. This sequential focusing is the reason they move their head from side to side, and that head movement makes them zigzag when they move. Having eyes on either side of the head gives chickens a better all-around sight with which to spot their many enemies: always-hungry dogs, cats, raccoons, hawks, weasels, possums, foxes, and rats. Their eyesight is nevertheless keen, and they can spot minute objects while hunting for food.

Chicken Facts

The heartbeat of chickens is fast, occurring about 286 times a minute in males and 312 in females in a resting condition.

The chicken's temperature is about 107 to 107.5 degrees Fahrenheit.

Hens start laying eggs at about six months. In commercial egg production, birds are about a year and a half when they are replaced by new, young egg layers.

A cluster of chicks is depicted in this romantic nineteenth-century print by famed German-American lithographer Louis Prang. (Courtesy of the Library of Congress)

The Edible Chicken

- A broiler or fryer usually weighs between two-and-a-half and three-and-a-half pounds.
- A roaster weighs four pounds or more. It is a young meaty chicken that can be cooked tender by roasting.
- A stewing chicken is a mature hen, often over the hill after a life of egg production. It will need to be stewed to get to a tender state.

Practitioners of voodoo sacrifice chickens at ceremonial feasts for the gods. They believe that birds symbolize a link to the other world.

Israeli researchers have developed a featherless chicken that is claimed to be better for the environment (because there is no feather waste) but needs to be protected from the sun to prevent sunburn.

A hen has 340 taste buds, will eat just about anything, and loves carbohydrates like cooked rice, bread, and corn.

Chickens can differentiate color and, in a test, preferred red.

Chickens have comparatively short life spans. Some live to be ten to fifteen years old, but they are the exception. Chicken expert Barry Koffler from www.feathersite.com has determined that hens that live past four years without reproductive system failures live on an average of eight to eleven years. Roosters average five to seven years.

SOIGNONS LA
BASSE-COUR

G. Douanne
16 ans

VILLE de PARIS Ecole de Filles Avenue Daumesnil

JE SUIS UNE BRAVE POULE DE GUERRE
JE MANGE PEU ET PRODUIS BEAUCOUP

UNION FRANÇAISE 286. Boulevard St Germain. PARIS.

Comité National de Prévoyance et d'Economi

CETTE AFFICHE NE DOIT PAS ÊTRE VENDUE.

Chapter 4

Eggs and Chicks

The color of the eggshell bears no relation to the quality or taste of the egg. Nor does it indicate whether the egg is organic, meaning the hen is fed a strictly organic diet; free range, which means the hen had access to grazing outdoors; or from an "egg factory," a mass–egg-laying operation. Some purchasers like white eggs, some prefer brown, and some like the green or blue coloring. It is merely a personal preference. But how does the egg get its color? In simple terms, some chickens secrete a brownish pigment into the outer layers of the shell in the oviduct where eggs are formed, and others do not.

Female chicks, or pullets, are born with thousands of cells in their ovary, which produces tiny unfertilized ovum. By the way, while hens have two ovaries, only one, normally the left, is functional. When the hen has reached maturity, one egg cell is released each day and drops down the oviduct, where it can be fertilized by a rooster. Layers of white are produced around the yolk, a membrane develops, and finally the shell is added. Eggshells are made of calcium carbonate, secreted by tiny glands in the oviduct. The hen's body takes from eighteen to thirty-six hours to produce an egg, regardless of whether or not it is fertilized. Typical store-bought eggs are not fertilized, so attempting to hatch them would not yield the desired results.

Eggs are extremely nutritious. A valuable source of protein, each egg contains six to seven grams of the daily protein requirement. Calorie counters will be glad to note that one egg

Above: A chick used its egg tooth to cut the egg membrane and hatch itself out of the shell. (Courtesy of Diane Fadus)

Facing Page: "I am a fine war hen. I eat little and produce a lot," reads this French World War I lithograph. Certain breeds have long been prized for their hens' prolific egg-laying, including the Crevecoeur. (Courtesy of the Library of Congress)

contains only about eighty calories. Oleic acid is the main unsaturated fat and has no effect on blood cholesterol. Eggs are a good source of vitamin A; the B vitamins thiamin, riboflavin, and niacin; and vitamin D, along with minerals such as iron and phosphorus. But don't count on eggs for your daily supply of calcium (it's all in the shell) or vitamin C. They also contain few carbohydrates and no dietary fiber.

If an egg is fertilized by the rooster and incubated by either a hen or an artificial incubator, it can develop into a baby chicken. A fertilized egg has a tiny white disc (1/25 inch in diameter) on top of the yolk. Within three days, this disc has developed a head, body, tail, and heart, augmented by a blood supply. By the tenth day, the legs, wings, toes, and the beginning of feathers are formed. A few days before it is ready to hatch, the chick can be heard peeping inside the shell. The chick then starts breaking its way out of the shell with its egg tooth, a special, sharp nail at the end of the beak that drops off after hatching. An egg can take up to fourteen hours to hatch as the chick makes a rough line around the middle of the egg before finally pushing the shell apart. So, basically, if an egg is incubated, in twenty-one days, voilà, a chicken!

Chickens weigh about one-and-a-half ounces at birth. They are wet when born and can stand on their first day after they have dried off. There is nothing more adorable, in my opinion, than baby chickens. Newborn, they are little balls of fluff, immediately busy, scurrying around, pecking, peeping, and keeping a nervous eye on their mothers.

Identifying Chick Breeds

Certain chicken breeds are born with identifying features:

- Ancona chicks have patches of black and white.
- Blue Andalusian chicks have a bluish tint at hatching.
- White Australorp chicks have velvety black spots on the head and back.
- Light Brahma chicks are white with feathered legs.
- Buff Cochin chicks have feathered legs, and their color changes very little from hatching.
- Jersey Black Giant chicks are black with quite a bit of white at hatching.
- White Leghorn chicks are yellow at hatching and have yellow legs.
- Barred Plymouth Rock chicks are mostly black at hatching with a few specks of white.
- Polish chicks have a crest.
- Rhode Island Red chicks are red with yellow markings.
- White Rock chicks are either yellowish white or slaty white.

Sexing Chicks

Because newborn chicks have been living off the yolk, using the white for nutrition, they do not need food or water for the first forty-eight hours. This makes shipping day-old chicks feasible. I have often arrived at the post office to pick up a small box of peeping chicks, amazing bystanders.

But before shipping off a box of chicks, the breeder must sex the chicks.

A pair of Buff Brahma chicks stay warm on their mother's back. (Photograph © Alan and Sandy Carey)

Identifying the sex of a chicken at birth is important for buyers who would prefer egg layers to roosters. Since the major sex characteristics, like combs, wattles, and tail feathers, aren't apparent for another five to six weeks, breeders must use other methods to immediately identify whether the chick is a he or a she. There are two ways to identify the sex of the chicks at birth: use the Japanese method or know the traits characteristic of crossbreeds.

Japanese scientists developed a method of sexing chicks in 1924. Trained chicken sexers first arrived in the United States in 1933 and fetched a pretty penny (more than four thousand dollars a day) for their keen ability to sex five to seven thousand chicks a day. The technique consists of inverting the chick and studying the cloacae with either the naked eye or a binocular eye loupe.

Another method involves crossbreeding. In some crossbreeds, the sex of the chick determines its color or feathering. These sex-linked traits easily tell a breeder which chicks are cockerels and which are pullets. For example, when a Delaware hen is crossed with a New Hampshire or Rhode Island Red rooster, the male chicks have the Delaware pattern of the mother and the female chicks have the solid red feather pattern of the father. Another example is the sex-linked offspring of certain roosters, such as the Brown Leghorn or Rhode Island Red, crossed with the Barred Plymouth Rock hen. The male chicks have a white patch on top of the head with yellow beaks and shanks, and the female chicks have black heads with dark beaks and shanks.

Chapter 5

Everything But the Cluck

Chickens are not merely used in stew, pot pies, or casseroles, and their eggs are valuable for more than tasty omelets. Every ounce of these fabulous fowl can be used for many purposes. Other animals enjoy the good flavor of fowl as a food item as much as humans. Many pet-food companies, including Purina and IAMS, use chicken as a key ingredient for their natural-protein products.

New benefits from chickens and their components are constantly being researched. Anna Edey of Solviva, Inc. has invented solar dynamic, bio-benign greenhouses that use live chickens as an additional source of heat. According to her calculations, each chicken emits about eight British thermal units (Btus) of energy per pound per hour, which gives the heat equivalent to two-and-a-half gallons of fuel oil per animal for six months' heating. Her showcase facility is on Martha's Vineyard in Massachusetts, and she sells plans for similar units.

Facing Page: The Red Dorking has a large single comb. Rooster combs are a source of hyaluronan, a gelatinous substance used to lubricate and protect the eye during ophthalmologic surgery on animals. (Photograph © Lynn M. Stone)

Feathers

What can be done with the more than two billion pounds of chicken feathers produced each year in the United States? Farmers usually throw them away or grind them up to add protein to animal feed. From our home flock, we add small feathers to enrich our compost pile or donate the more colorful ones to schools or individuals for art projects.

Others are more inventive. Walter Schmidt and other chemists with the U.S. Agricultural Research Service have been studying various uses of chicken feathers for years. They combine chicken feathers with fiberglass to make boat exteriors and automobile parts such as dashboards, interior panels, and parts of the glove compartment. Using a feather base for these articles makes a vehicle lighter and more economical, resulting in less fuel consumption.

Schmidt and other researchers are also experimenting with using chicken feathers to produce a material called feather fiber, which can be used in place of wood or wood pulp. Feather fiber is finer than wood pulp and, when used as an air filter, could collect more spores, dust, and dander, thus dramatically improving the air quality in homes and offices.

Scientists are also developing a form of paper from chicken feathers. Prototypes have resulted in paper with unusual textures and dyeing properties that enable it to be reused many times.

Richard Wool and chemical engineers from the University of Delaware are working on a new generation of computer microchips that replace silicon with a material made from chicken feathers. Tests show that electrical signals moved twice as quickly through the chicken-feather chip as through a conventional silicon chip. The use of chicken feathers makes the chips cheaper, as well as more environmentally friendly when disposed of.

A beautiful Cree rooster shows off the long hackle and saddle feathers prized by anglers for tying flies. (Photograph © Alan and Sandy Carey)

In the fly-fishing industry, Tom Whiting and others raise specialty chickens with long hackles to use in custom fly-tying applications. Whiting Farms, Inc., in Colorado, breeds chickens with a wide, webby neck hackle that can be used for bass bugs and saltwater flies; genetically engineered saddle hackles perfect for tiny dry flies; and even hackles that can be used specifically for realistic mayfly tails. He also uses feathers to make beautiful stuffed and mounted chickens, feather pottery, and other decorative articles.

In the Philippines, the Pililla Poultry Processing Plant has developed technology that can convert chicken feathers into fertilizer. It involves treating the feathers with an enzyme-like product that liquefies not only the plumage but also the entire chicken carcass. The resulting nitrogen and phosphorus-rich liquid can be used as an organic fertilizer.

Combs

Rooster combs are one of the world's richest sources of hyaluronan, formerly known as hyaluronic acid. Hyaluronan has many medicinal uses, such as protecting the eye during veterinary eye surgery, reducing inflammation in arthritic knees of racehorses, and preventing post-surgery scar tissue.

Hyaluronan was discovered in 1934 by Karl Meyer in an ophthalmology lab at Columbia University. Working with the viscous substance in cow eyeballs, he determined that the material helped the eye retain its shape. Later, in the early 1940s, also at Columbia, Endre Balazs figured out how to extract and purify hyaluronan from rooster combs. Today, it has become the latest treatment for plumping up human facial wrinkles! Additionally, physicians are using injections of hyaluronan to treat osteoarthritis of the knee.

Whiting Farms uses feathers from roosters, selectively bred for length and consistency of feathers, to tie fishing flies. The saddle feathers pictured here are the two most important colors and patterns used in dry flies: brown and "grizzly" (barred). (Courtesy of Thomas Whiting, Whiting Farms, Inc.)

Glossary

bantam: A miniature chicken. Some, such as Dutch Bantams and Rose Comb Bantams, are distinct breeds and considered "true" bantams; others have been bred as miniatures of large breeds and are approximately one-fourth to one-fifth the size of their large counterpart.

beard: A beard-like cluster of feathers below the beak. Certain fowl, such as Ameraucanas, Bearded Polish, Belgian Bearded d'Anvers, Belgian Bearded d'Uccles, Crevecoeurs, Houdans, Faverolles, Bearded Silkies, and Sultans are described as bearded.

blade: The portion of the males' single comb that is to the rear of the base.

booted: Feathered shanks and toes and vulture hocks. This characteristic is seen on Booted Bantams, Sultans, and Belgian Bearded d'Uccles.

Booted

A White Ameraucana sports the beard and muff distinctive to the breed. (Photograph © Alan and Sandy Carey)

broody: Wanting to sit on and hatch eggs: *a broody hen.*

capon: A castrated rooster, often fattened for eating. To castrate a rooster, a small incision is made under the wing between the last two ribs, and the testicles, attached to the spinal column at the rear of the lungs, are removed.

carriage: The attitude and style a bird demonstrates in shows. It is the way an individual bird carries itself as it moves normally.

chick: The young of domestic hens. Also a cute girl.

Facing Page: The Polish is famed for its striking crest. (Photograph © Alan and Sandy Carey)

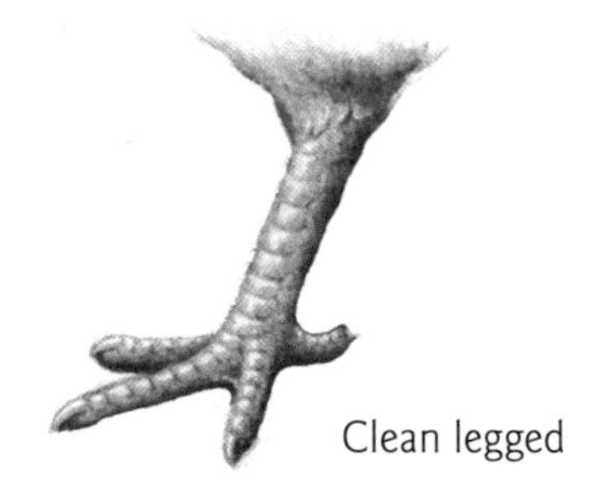
Clean legged

clean legged: Having no feathers on the shanks.

cock: An adult male domestic fowl, also known as a rooster.

cockerel: A young male domestic fowl, usually under a year.

coverts: The feathers covering the base of the main tail feathers in males and most of the tail feathers in females. The male's are curved and the female's are oval.

crest: A tuft of feathers growing from a bony knob on the top of the head. A rounded crest is found on the Polish, Houdan, Crevecoeur, Silkie, and Sultans.

down: The fluffy part of the feather near the base and the soft fluff on newly hatched chicks.

dual-purpose: A chicken bred for the dual purpose of laying many eggs and providing high-quality meat.

dub: To trim the comb, wattles, and earlobes for show purposes. Modern and Old English Game breeds are dubbed for exhibition. Originally, roosters were dubbed for cockfighting, allowing less surface area for the opponent to attack.

ear: Chicken ears are located above the earlobes and covered with plumage.

earlobe: Visible patches of bare skin in various sizes and colors, including red, white, blue, or purple.

ear tufts or **ear whiskers**: Feathers that grow straight out from the earlobe region. Some even grow in a curve to form a circle about the bird's heads. Ear tufts are characteristic of the Araucana.

face: The smooth skin around and below the eyes. The color is usually red, white, or purple.

feather legged: Having shank feathering—feathers on the outer side of the shanks and outer and/or middle toes. Cochins and Brahmas are feather-legged breeds.

Feather legged

fluff: Soft downy feathers running from the abdomen and to the rear of the thighs.

fowl: A domestic, granivorous bird of any kind, including chickens, guinea fowl, ducks, geese, peacocks, and turkeys.

frizzle: Feathers that curl and curve outward and forward. This genetic variation may occur in any breed.

Gallus domesticus: The common domestic chicken.

Gallus gallus: The red jungle fowl, the ancestor of the chicken, also called *Gallus Bankiva*.

gamecock: A fighting rooster.

game fowl: Any of several breeds generally used in cockfighting.

gamy tail: A tail that is slim, tightly folded, and tapered at the end. It is typical of the Modern Games, Cornish, and some Malays.

gizzard: An internal organ in birds, often containing grit, that aids in the grinding of food.

grit: Small pebbles that, when eaten by birds, aid in digestion by grinding the food in the gizzard.

hackles: The feathers on the side and rear of the chicken's neck plumage. Male hackles are more pointed than those of the female, except in breeds having hen-feathered males.

hard feather: The plumage of game fowl. The feathers are narrower and shorter than those of other chickens, and the shaft and barbs are tougher and closely knit. There is little fluff.

hen: A mature female of common domestic fowl.

hen-feathered: The feathering of males that do not develop the long sickle feathers, pointed hackle feathers, or pointed saddle feathers that are typical of a rooster. Sebright Bantams must be hen-feathered to fit the breed standard, and Campines exhibit a modified version of hen-feathering.

hock: The joint between the lower thigh and shank, sometimes incorrectly referred to as the knee.

keel: The lower edge of the breastbone in chickens, resembling a ship's keel.

leader: See **spike**.

Spike or Leader

lopped comb: A comb that falls to one side. A lopped comb looks like a "comb-over" hairstyle. It is typical of many of the hens of Mediterranean breeds.

molt: The seasonal shedding of a fowl's feathers.

muff: A cluster of feathers below and around the sides of the eyes, covering the earlobes. Also known as whiskers, they are found in combination with a beard.

pearl eye: A creamy-colored or pale bluish gray eye, characteristic of the Cornish, Malay, Shamo, and Aseel.

pecking order: The social rank in a flock of chickens in which dominant birds peck subordinate birds.

poultry: Domesticated birds raised for meat, eggs, or pets, including chickens, ducks, turkeys, and geese.

preen: The act of grooming and cleaning the feathers done by fowl.

pullet: A female chicken under a year of age.

rooster: A mature male chicken, often referred to as a "cock."

saddle: The part of the chicken to the rear of the back, before the covert feathers. In the male, it is covered with long pointed feathers called saddle feathers. There are both upper and lower saddle feathers.

scales: The thin, overlapping plates completely covering the shanks and top of the toes of a fowl.

sex feather: A feather in the hackle, back, saddle, and sickles. The rooster's are typically pointed and the hen's rounded.

sex-linked: Sex-linked chickens are planned crosses that can be sexed at hatching and are hardy and often more productive than their parents' respective breeds. The sexually determined inherited trait is usually color in chicks, but it can also sometimes be feather development.

shank: The portion of the leg between the foot and the hock.

shank feathering: See **feather legged**.

sickles: The long curving feathers in a rooster's tail. There are two types of sickles: greater (or main) and lesser. The greater sickles are the prominent middle uppermost pair of tail feathers. The lesser sickles hang to the side and cover most of the main tail.

silkie feathers: Feathers that have thin shafts and long barbs that do not web, making the feathers soft and fluffy. This feather type is characteristic of Silkies.

spike: The pointed end of a rose comb, sometimes also called the "leader."

spur: A rounded or pointed, stiff, horny projection on the rear inner side of the shanks, prominent in the male and occasionally found in females. Also a metal spike attached to a gamecock's leg.

squirrel tail: A tail in which any portion projects forward of the vertical from its anterior base, as found in Japanese Bantams. It is a disqualification in most other breeds.

standard: The ideal specimen of a chicken in a breed, set forth by the *American Standard of Perfection*.

stern: See **fluff**.

strain: A group of related birds within a breed that have been bred as a closed population for a number of years and reproduce uniform characteristics with marked regularity.

tail feathers: The main straight, stiff, long feathers of the tail, located under and between the hen's covert feathers and the rooster's covert and sickle feathers.

under color: The color of the lower fluff portion of the feather, not visible when feathers are in their natural position.

vulture hocks: Stiff, long feathers growing out of the lower thigh and projecting downward and backward. They are seen in Belgian Bearded d'Uccles, Booted Bantams, and Sultans.

wattle: A thin pendulous growth of flesh that dangles from the neck and throat, larger in males than females.

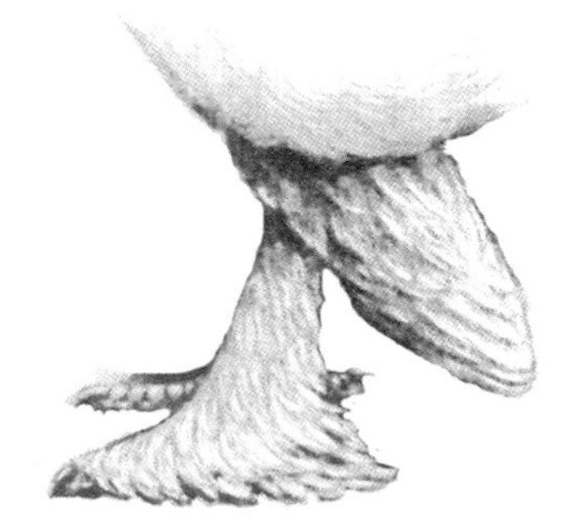

Vulture hocks

American Poultry Association
Breed Classification Table

Large

American
Buckeye
Chantecler
Delaware
Dominique
Holland
Java
Jersey Giant
Lamona
New Hampshire
Plymouth Rock
Rhode Island Red
Rhode Island White
Wyandotte

Asiastic
Brahma
Cochin
Langshan

English
Australorp
Cornish
Dorking
Orpington
Redcap
Sussex

Mediterranean
Ancona
Andalusian
Catalana
Leghorn
Minorca
Sicilian Buttercup
Spanish

Continental (North European)
Barnevelder
Campine
Hamburg
Lakenvelder
Welsummer

Continental (Polish)
Polish

Continental (French)
Crevecoeur
Faverolles
Houdan
La Flèche

All Other Standard Breeds
Ameraucana
Araucana
Aseel
Cubalaya
Frizzle
Malay
Modern Game
Naked Neck
Old English Game
Phoenix
Shamo
Sultan
Sumatra
Yokohama

Bantam

Game

Modern Game
Old English Game

Single Comb Clean Legged Other Than Game

Ancona
Andalusian
Australorp
Campine
Catalana
Delaware
Dorking
Dutch
Frizzle
Holland
Japanese
Java
Jersey Giant
Lakenvelder
Lamona
Leghorn
Minorca
Naked Neck
New Hampshire
Orpington
Phoenix
Plymouth Rock
Rhode Island Red
Spanish
Sussex
Welsummer

Rose Comb Clean Legged

Ancona
Belgian Bearded d'Anvers
Dominique
Dorking
Hamburg
Leghorn
Minorca
Redcap
Rhode Island Red
Rhode Island White
Rose Comb
Sebright
Wyandotte

All Other Combs Clean Legged

Ameraucana
Araucana
Buckeye
Chantecler
Cornish
Crevecoeur
Cubalaya
Houdan
La Flèche
Malay
Polish
Shamo
Sicilian Buttercup
Sumatra
Yokohama

Feather Legged

Belgian Bearded d'Uccle
Booted
Brahma
Cochin
Faverolles
Frizzle
Langshan
Silkie
Sultan

How to Use the Breed Profiles

There are hundreds of chicken breeds and thousands of varieties throughout the world. As a result, serious chicken choices had to be made when writing this book, and I hope no chicken's feathers will be fluffed out of joint because of omissions. Because of space limitations, the breed profiles are limited to the breeds itemized in the *American Standard of Perfection* and recognized by the American Poultry Association.

Every chicken falls into one of two major breed categories: large and bantam. Large chickens and their smaller bantam counterparts have similar traits, although bantams are not exact miniatures. Bantams have different proportions than large chickens; the bantams' heads, wings, tails, and feathers are disproportionately large for their body size. The choice of raising large or bantam chickens is a matter of personal preference. Exhibitors of large chickens appreciate the size, grandeur, and often majestic quality of their birds. Farmers rely on large chickens for a source of meat and eggs, and large chickens are the mainstay of the poultry industry.

The word "bantam" (after Bantam, a province in Java) refers to the more than 350 kinds of smaller chickens. They exist in almost every variety seen in large chickens. In addition, some "true" bantams, including Antwerp Belgians, Booted Bantams, Dutch Bantams, Japanese Bantams, Rose Comb Bantams, and Sebrights, have no large counterpart. Silkies are considered true bantams in the *American Standard*, but the breed is divided into two classes in Europe: Silkie bantams, which are smaller than ours, and large Silkies.

Bantams are raised primarily for their pulchritude, for exhibition, and as pets. The advantages of raising bantams are that they take less space, eat less, transport more easily, and lay just as many eggs as large chickens. Although their eggs are smaller than those of their larger relatives, they are just as delicious. Bantams have been present in the United States since the early 1700s, but their popularity did not take off until the early 1900s. In 1914, the American Bantam Association formed independent of the American Poultry Association.

This guide profiles each breed, which in turn belongs to a class. Within each breed, there are varieties.

Facing Page: The Black-Tailed White Japanese Bantam is one of the many popular bantam varieties raised for exhibition and as pets. (Photograph © Lynn M. Stone)

Breed

A breed is a group of chickens that share ancestors and similar physical features. These characteristics include a common body shape or type, skin color, egg color, number of toes, and shank feathering or non-feathering. For the most part, chickens within a breed have similar combs, although there are exceptions.

Status

Many chicken breeds are in danger of extinction. These breeds have been overlooked as large-scale poultry production soared in response to the growing human population and popularity of chicken as a food source. Only a few lines of chickens are used commercially for eggs and meat, excluding many of the original breeds once important in the chicken industry.

The American Livestock Breeds Conservancy (ALBC) has been instrumental in creating awareness of endangered chicken breeds. The breeds that need attention have been pinpointed thanks to the study "Identifying Breeds in Danger of Extinction" by noted American researchers Marjorie E. F. Bender, Robert O. Hawes, and Donald E. Bixby. Alarmingly, thirty-four chicken breeds, or nearly half the historic breeds, are reported to be endangered.

The dangers are indicated in the profiles in the following ways:

critical: Fewer than five hundred breeding birds and five or fewer primary breeding flocks in North America.

rare: Fewer than one thousand breeding birds and seven or fewer primary breeding flocks in North America.

watch: Fewer than five thousand breeding birds and ten or fewer primary breeding flocks in North America. Also included are breeds that present genetic or numerical concerns or have a limited geographic distribution.

study: Breeds that are of interest but lack either definition or genetic or historical documentation.

recovering: Breeds that were once listed in one of the other categories and have exceeded watch category numbers but still need monitoring.

The status category is omitted from the profiles for breeds that are not considered endangered.

The Society for the Preservation of Poultry Antiquities (SPPA) is another organization active in preserving rare and endangered breeds of chickens. Established in 1967, the SPPA aims to perpetuate and improve rare breeds of poultry and to help locate rare stock. An SPPA's breeders directory offers a list of the hundreds of breeders of endangered poultry.

Australorps are listed as recovering by the American Livestock Breeds Conservancy, along with Orpingtons, Plymouth Rocks, Rhode Island Reds, and Wyandottes. (Photograph © Lynn M. Stone)

Class

With large chickens, the class generally designates the area where they originated. The term "originated" refers to when the breed originally came into being, while "developed" refers to when breeders consciously worked to evolve the birds. Sometimes these are the same; sometimes they are different. For example, the Araucana originated in South America but was developed by breeders in England and the United States. With many breeds, it's impossible to trace the ancestry farther back than the point of development, but with some ancient breeds, like the Aseel, we can follow the lineage back thousands of years, long before modern breeding records.

The classes of large fowl are American, Asiatic, English, Mediterranean, Continental (including North European, Polish, and French), and All Other Standard Breeds (including Orientals and Miscellaneous).

Bantam classes are divided differently. Bantams are earmarked for show purposes by the type of comb, whether or not they have feathered legs, and by the game varieties. Bantam classes are Game Bantam, Single Comb Clean Legged (Other Than Game Bantams), Rose Comb Clean Legged, All Other Combs Clean Legged, and Feather Legged.

Varieties

Varieties are subdivisions of a breed. The body shape is identical within a variety. The varieties differ in the color of plumage, type of comb, and whether or not the chicken has a beard and muff. In the profiles, the dates listed in parentheses behind the varieties signify the years those particular varieties were admitted to the *Standard*.

Weight

The size of the chicken is important in its identification. Large chickens range from about three to thirteen pounds. Hens weighs less than roosters. Bantams weigh one-fourth to one-fifth of their large counterpart, generally 16 to 30 ounces. The profiles give the *Standard*'s weight for the large cock, hen, cockerel, and pullet. Except for true bantams, the bantam's weight requires a calculator.

Comb

The comb is an important characteristic in identifying chickens. The type of comb, its size, and its color all factor into the determination of a breed. Some combs, including the buttercup, carnation, cushion, pea, strawberry, and V-shaped, have little variation. The rose and single combs, however, may have variations within the breeds. For example, the tapering spike at the end of the rose comb may vary. Some turn upward, others are horizontal, and others follow the contour of the head. The single comb may have five or six serrated points and vary in size depending on the breed. The female's single comb may be erect or lopped.

Tail

The tail shape, size, color, and angle of carriage all assist in chicken identification. Some breeds, such as Rhode Island Reds and Whites, carry their tails as low as 20 degrees above the horizontal, while others, such as Booted Bantams and Belgian Bearded d'Uccles, carry theirs as high as 75 degrees. In general, the rooster carries his tail higher than the hen does. For example, the Chantecler hen carries her tail at 30 degrees, the rooster at 45 degrees. A few ancient breeds—including the Aseel, Cornish, and Malay—carry their tails below the horizontal. The long-tailed Sumatra and Cubalaya carry their exotic tails horizontally, and the ends of their feathers drag on the ground. Short, round tails, such as the Silkie's and the Cochin's, are not considered to have an angle.

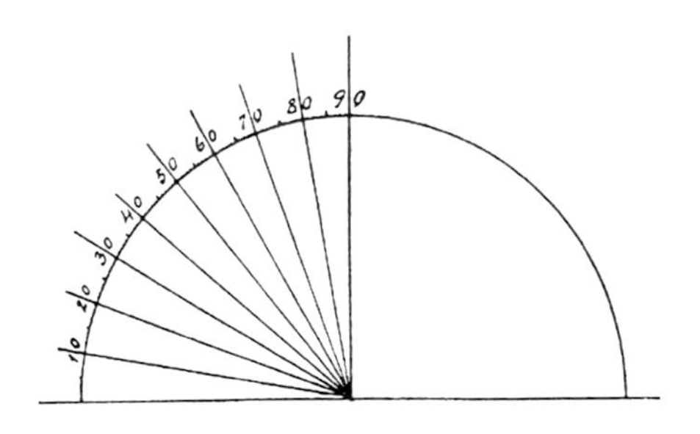

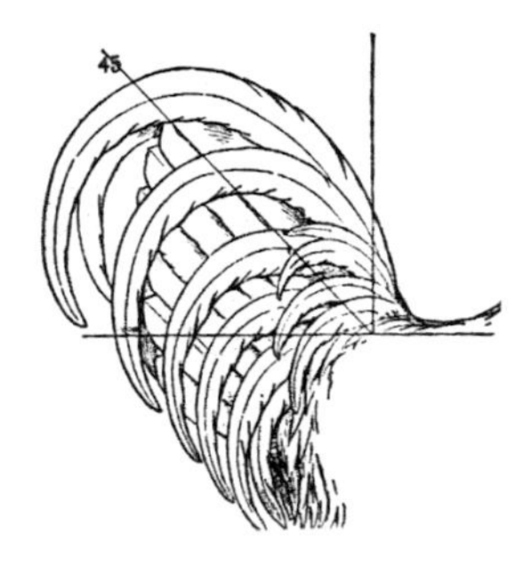

Chicken tails are measured by their angle of carriage.

Legs

To make things simple, let's understand *legs* to mean both legs and feet. Chickens have two legs, and the feet and shank portions of the legs are covered with scales. The color of the legs ranges from shades of yellow to willow, dark slate, and even black. Their texture ranges from smooth to either partially or totally feathered. Some breeds also have vulture hocks.

Eggs

Eggs come in a variety of colors and sizes. The color of the eggshell is not related to the diet of a chicken. According to Richard Balander, associate professor of animal science at Michigan State University, the color of the earlobe correlates to the color of the egg. He claims that breeds with white earlobes lay white eggs (with the exception of the Penedesencas from Spain, which lays dark chocolate-colored eggs) and chickens with red earlobes lay brown eggs. There are several exceptions in the *Standard* breeds. The Holland and Lamona have red earlobes but lay white eggs, and Araucanas have bright red earlobes but lay green-blue eggs.

Eggs range in size from tiny to large. Typically, bantams lay smaller eggs and large breed chickens lay larger eggs.

Other

To help with chicken identification, the profiles list any other distinguishing characteristics, such as distinctive plumage, unusual beards and muffs, or extraordinary differences between a breed's cock and hen.

Breed Profiles

Ameraucana

The Ameraucana was created in the 1970s in the United States. It was developed from the Araucana, which is well known for laying "Easter eggs" with shells of beautiful shades of blue and green. The Ameraucana retains the Araucana's characteristic eggs but has an improved meat quality. The Ameraucana differs from the Araucana in that it possesses a tail, has a beard and muff instead of ear tufts, and is larger and heavier.

Large

Class: All Other Standard Breeds (Miscellaneous)
Varieties: Black, Blue, Blue Wheaton, Brown Red, Buff, Silver, Wheaten, White (1984)
Weight:
Cock: 6.5 pounds
Hen: 5.5 pounds
Cockerel: 5.5 pounds
Pullet: 4.5 pounds

Bantam

Class: All Other Combs Clean Legged Bantams
Varieties: Black, Blue, Blue Wheaten, Brown Red, Buff, Silver, Wheaten, White (1984)

Characteristics

Comb: Small pea
Tail: Medium length, the male's is carried at a 45-degree angle and the female's at 40 degrees
Legs: Slate-colored shanks and toes
Eggs: Medium, in shades of blue, green, and turquoise
Other: A full, well-rounded muff and beard, forming three separate lobes. Wattles are very small or absent.

Blue Wheaten Ameraucanas (Courtesy of William A. Suys, Jr.)

Ancona

The Ancona was developed in Italy and named after the port city of Ancona, located not far from Rome. The original name was Marchegiana, from the region of Ancona called The Marches, but that was a mouthful for the English-speaking people, so it was changed. From Italy, the bird was imported to England around 1848 and developed by fanciers. The Single Comb variety was developed first and, later, the Rose. Known primarily for producing large numbers of white eggs, Anconas became popular throughout Europe. They arrived in America in 1890. Bred from Leghorns and other fowl, Anconas bear a striking resemblance to Leghorns with their mottled coloration. Anconas are excellent egg layers and are non-sitters.

Single Comb Anconas

Large

Class: Mediterranean
Varieties: Single Comb (1898), Rose Comb (1914)
Weight:
Cock: 6 pounds
Hen: 4.5 pounds
Cockerel: 5 pounds
Pullet: 4 pounds

Bantam

Class: Single Comb Clean Legged Other Than Game Bantams
Varieties: Single Comb (1960), Rose Comb (1960)

Characteristics

Combs: Medium single or medium rose. The single is straight with five distinct points; the female's droops over after the first erect point. The rose is low and square in front and tapers to a horizontal spike.
Tail: Large, the male's is carried at a 40-degree angle and the female's at 30 degrees.
Legs: Moderately long, yellow shanks and toes
Eggs: Small, white
Other: The lustrous greenish black plumage is mottled, and each feather has V-shaped white tips.

Status: Rare

Andalusian

The Andalusian, one of the oldest of the Mediterranean breeds, is named after Andalusia, the province in Spain where it is thought to have originated. Developed in 1846, the breed is closely related to the Black Spanish. There is only one variety accepted in the *American Standard*, the Blue Andalusian, which has a bluish slate coloration. However, when two Blues mate, some chicks come out blue, some black, and some splash (or white). The Blue variety is produced by mating a black and splash. This breed was used in Gregor Mendel's experiments of heredity and genetics.

Large

Class: Mediterranean
Variety: Blue (1874)
Weight:
Cock: 7 pounds
Hen: 5.5 pounds
Cockerel: 6 pounds
Pullet: 4.5 pounds

Bantam

Class: Single Comb Clean Legged Other Than Game Bantams
Variety: Blue (1960)

Characteristics

Comb: Medium single. The male's is erect with five well-defined points. On the female, the first point is erect and the remainder gradually droop to one side.
Legs: Long, dark slate blue shanks and toes
Tail: Long and full, the male's is carried at a 40-degree angle and the female's at 30 degrees.
Eggs: Large, chalk white
Other: The feathers are edged with a dark, almost black, shade. The male's plumage has a rich black overcast with a bluish tinge; the overall cover of the female's is more of a bluish slate.

Status: Critical

Blue Andalusians

Araucana

Raised by Araucana Indians in the Arauca region of central Chile, this breed was discovered by missionaries in the 1500s. Ferdinand Magellan in 1519 described poultry that resembled Araucanas. It is believed that the Araucana descends from a wild jungle fowl that was domesticated perhaps from a type of pheasant called the "chachalaca." Araucanas still exist in the wild in the Amazon Basin and pockets of the Andes Range.

The breed was later developed in Great Britain and the United States. According to the British Araucana Society, the Araucana first arrived in England in the early 1920s and was imported to the United States in the early 1930s. Also in the '30s, Scottish breeder George Malcolm created a beautiful lavender variety. Interestingly, the *United Kingdom Standard* calls for a small crest and tail, while Araucanas in the *American Standard* are rumples, meaning that both male and female do not possess a tail. In fact, the entire coccyx is missing.

The Araucana's popularity stems from its beautiful blue-green eggs. Its hybrids lay eggs in a variety of colors, including blue, pale green, tan, turquoise, olive, and even pinkish.

Large

Class: All Other Standard Breeds (Miscellaneous)
Varieties: Black, Black Red, Golden Duckwing, Silver Duckwing, White (1976)
Weight:
Cock: 5 pounds
Hen: 4 pounds
Cockerel: 4 pounds
Pullet: 3.5 pounds

Bantam

Class: All Other Combs Clean Legged Bantams
Varieties: Black, Black Red, Golden Duckwing, Silver Duckwing, White (1976)

Characteristics

Comb: Pea
Tail: Absent
Legs: Willow shanks and toes, except for the White variety, which has yellow
Eggs: Medium, in shades of blue
Other: Araucanas have ear whiskers or tufts that grow from the earlobe region in all different directions. Their earlobes are red, and their wattles are small or absent.

Status: Study

Araucanas

Aseel

The Aseel, or Asil, is thought to be the oldest breed of game fowl, bred in India for cockfighting more than five thousand years ago. *Aseel* is a Hindu adjective meaning "high born" or "aristocratic." These birds were kept by the princes of India, who valued their pugnacious qualities. Chicken historian Lewis Wright says, "There can be little doubt that the birds whose battles are alluded to in the Institutes of Menu, 1000 B.C. if not the Aseel as now known, were at least their ancestors." Although the cocks are aggressive fighters with the reputation of troublemakers when confined, they are said to be docile when kept apart from other cocks and one of the most intelligent of all breeds. Although broody, the Aseel hen may lay only two to three eggs a year. The birds were once popular for their abundant supply of breast meat.

Spangled and Black Breasted Red Aseels

Large

Class: All Other Standard Breeds (Orientals)

Varieties: Black Breasted Red, Dark, Spangled, White (1981), Wheaten female (1996), Wheaten male (2000)

Weight:

Cock: 5.5 pounds

Hen: 4 pounds

Cockerel: 4.5 pounds

Pullet: 3 pounds

Bantam

There are no Aseel bantams in the *Standard*.

Characteristics

Comb: Pea

Tail: Medium length, carried below the horizontal

Legs: Yellow shanks and toes. The legs are short and widely spaced apart.

Eggs: Small, white or tinted light brown

Other: They have a small head, devoid of feathers, and a short, strong beak. Their plumage is hard, close, wiry, and devoid of fluff.

Status: Critical

Australorp

The Australorp descends from Black Orpingtons bred by English breeder William Cook in the late 1880s. The Orpington was exported to Australia and there developed for its egg production and as a good table fowl. Australians had great success creating this prolific layer, which boasts egg records as high as 339 in one year. At first, they were called Austral Orpingtons but the name was later abbreviated. The Australorp returned to England in 1921.

Large

Class: English
Variety: Black (1929)
Weight:
Cock: 8.5 pounds
Hen: 6.5 pounds
Cockerel: 7.5 pounds
Pullet: 5.5 pounds

Bantam

Class: Single Comb Clean Legged Other Than Game Bantams
Variety: Black (1960)

Characteristics

Comb: Single, erect with five distinct points, the first smaller
Tail: Moderately long tail, carried at a 40-degree angle
Legs: Shanks and toes are black when the birds are young and dark slate when they become adults. The toes and bottoms of the feet are pinkish white.
Eggs: Medium, tinted brown

Other: The plumage has a greenish black surface color and a black under color.

Status: Recovering

Black Australorps

Barnevelder

The Barnevelder is the most famous of the many breeds in Holland. It was developed in the district of Barneveld, Holland, where the Dutch Poultry Museum is located. Although Phoenician traders brought poultry to the area as early as the twelfth century, it was not until the mid-1800s that a conscious effort was made to improve egg production. Locals began mixing Barnevelders with exciting new Asiatic breeds, including Brahmas, Cochins, Langshans, and Malays. As a result, they developed a prolific layer of brown eggs, the choice color at that time. The Barnevelder made its exhibition debut in 1911 at The Hague's agriculture exposition and, in 1921, gained tremendous recognition at the first World Poultry Congress in The Hague. The *American Standard* recognizes only the Double Laced Partridge variety, although in Holland they also exhibit Blue Laced, White, and Black.

Double Laced Partridge Barnevelders

Large

Class: Continental (North European)
Variety: Double Laced Partridge (1991)
Weight:
Cock: 7 pounds
Hen: 6 pounds
Cockerel: 6 pounds
Pullet: 5 pounds

Bantam

There is no Barnevelder bantam in the *Standard*.

Characteristics

Comb: Medium single with five erect points
Tail: Full, carried at an angle up to 50 degrees
Legs: Yellow shanks and toes
Eggs: Large, dark brown
Other: The male's plumage varies much more in color than the female's. The male's head, neck, and saddle feathers are black with reddish brown edging and shafts and black tips; the back feathers and a portion of the wing are reddish brown with a wide lacing of a lustrous greenish black; and the tail is black with lustrous green coloring. In contrast, the female's plumage is an overall lustrous greenish black.

Belgian Bearded d'Anvers

A true bantam, the Belgian Bearded d'Anvers has a long history. Documentation began in the early seventeenth century and includes a painting by Dutch artist Albert Cuyp. Because the breed's city of origin has two names (Anvers in French, Antwerpen in Dutch), the chicken itself also has two names: Belgian Bearded d'Anvers and Antwerp Bearded Bantam. In American poultry circles, it is often referred to simply as the d'Anvers. In the nineteenth century, the breed is referenced in the French publication *Le Poullailler* ("The Chicken Coop") and numerous poultry exhibition records from Brussels. At the time, the Cuckoo and Black were the most common varieties. Brussels nineteenth-century breeder Michel van Gelder became fascinated with the fowl and searched throughout Europe for good breeding stock. He went on to develop the range of varieties in existence today.

Blue and Black Belgian Bearded d'Anvers (Courtesy of Heritage Poultry)

Bantam

Class: Rose Comb Clean Legged Bantams

Varieties: Black, Blue, Cuckoo, Mille Fleur, Mottled, Porcelain, Quail, White (1949), Self Blue (1981)

Weight:

Cock: 26 ounces
Hen: 22 ounces
Cockerel: 22 ounces
Pullet: 20 ounces

Characteristics

Comb: Rose with a spike that follows the contour of the head

Tail: Medium length and well spread, the male's is carried at a 75-degree angle and the female's at 65 degrees.

Legs: Dark slate or bluish slate shanks and toes in all varieties except the Cuckoo, which has bluish white legs. Legs are short and smooth.

Eggs: Tiny, creamy white

Other: The face is nearly covered by a muff and beard. The muff is thick and full and extends upward behind the eyes. The beard is full and connects with the muff.

Belgian Bearded d'Uccle

Porcelain Belgian Bearded d'Uccles

In the 1880s, breeders from the town of Uccle, Belgium, began developing the Belgian Bearded d'Uccle by crossing the Booted Bantam and the Belgian Bearded d'Anvers. The first varieties to be created were the Porcelain, White, and Mille Fleur. The last was the most popular because of its black tail and dark reddish brown feathers with white spangles. The d'Uccle was introduced to the American poultry scene around the turn of the century, and the Mille Fleur variety was admitted to the *American Standard* in 1914. Having no large fowl counterpart, the d'Uccle is one of the true bantams. According to the Belgian d'Uccle & Booted Bantam Club, poultry associations throughout the world recognize more than thirty varieties of the two breeds.

Bantam

Class: Feather Legged Bantams
Varieties: Bearded Mille Fleur (1914), Bearded Porcelain (1965), Bearded White (1981), Bearded Black, Bearded Golden Neck, Bearded Mottled, Bearded Self Blue (1996)
Weight:
Cock: 26 ounces
Hen: 22 ounces
Cockerel: 22 ounces
Pullet: 20 ounces

Characteristics

Comb: Single with five evenly serrated points
Tail: Long and erect, male's is carried at an angle of 75 degrees and female's at 70 degrees.
Legs: Feathered with vulture hocks and foot feathering
Eggs: Tiny, creamy white
Other: The Belgian Bearded d'Uccle is often compared to the Booted Bantam because of the breeds' similar body configuration, well-developed vulture hocks, and tail carried at a high angle. But the two breeds have distinct differences. The d'Uccle has a thick, full beard and muff that extend upward to eye level, while the Booted has no beard. The d'Uccle is also shorter and has a "bull," or fuller, neck and broader breast. And the d'Uccle has small or non-existent wattles, in contrast to the Booted's large wattles.

Booted Bantam

Originally from east Asia, the Booted Bantam was developed in Holland and is one of the oldest known Dutch poultry breeds. Captured on canvas by seventeenth-century painters such as Adriaen van Utrecht, the breed sports distinctive vulture hocks. A true bantam, its original name was Sabelpoot Kriel, which translates to "sword-legged bantam."

The Booted Bantam was imported to the United States and further refined by Massachusetts breeder E. C. Aldrich in the 1830s. The American version of the breed was accepted by the *Dutch Standard* in the early 1900s. Today in Holland, there are more than twenty varieties, the most popular being the Non-Bearded Mille Fleur.

Bantam

Class: Feather Legged Bantams

Varieties: Non-Bearded White (1879), Non-Bearded Mille Fleur (1914), Non-Bearded Porcelain (1965), Non-Bearded Black, Non-Bearded Self Blue (1996)

Weight:
Cock: 26 ounces
Hen: 22 ounces
Cockerel: 22 ounces
Pullet: 20 ounces

Characteristics

Comb: Medium single with five evenly serrated points

Tail: Long and erect, carried at an angle of 75 degrees

Legs: Feathered with vulture hocks and foot feathering

Eggs: Tiny, creamy white

Other: Large wattles and a U-shape between the neck and tail

Mille Fleur Booted Bantams

Brahma

The Brahma's ancestors hail from Asia's Brahmaputra River valley. There are several stories of how the breed came to exist in the United States. Some say the Brahma, a cross of Shanghais and Malays, was developed in Asia and then brought to America. C. N. Bement, author of *The American Poulterer's Companion*, supported this view, stating in the 1867 edition that a sailor brought the first Brahmas—three pairs, to be exact—to New York in 1850. He insisted that they arrived first in America and later in England. Others say that the breed was developed in the United States from imported Shanghais and Malays.

In any case, the Brahma was first exhibited in Boston in 1851 by Connecticut breeder O. S. Hatch. The first to develop the pea comb, Hatch considered it best for the New England climate because its small size resisted frostbite. George Burnham, of *The History of the Hen Fever* fame, was also active in the development of Brahmas. The name of the breed was officially change to Brahmapootra in 1851 and later shortened to Brahma. Brahmas became one of the leading Asiatic breeds in England, especially after Queen Victoria was given a quill pen made from a Brahma feather.

Large

Class: Asiatic
Varieties: Dark, Light (1874), Buff (1924)
Weight:
Cock: 12 pounds
Hen: 9.5 pounds
Cockerel: 10 pounds
Pullet: 8 pounds

Bantam

Class: Feather Legged Bantams
Varieties: Dark (1895), Light (1898), Buff (1946)

Characteristics

Comb: Small pea
Tail: Full, medium length, carried high and well spread
Legs: Well-feathered legs and feet, outside only
Eggs: Medium, brown
Other: The Light variety has Columbian plumage, with lustrous greenish black feathers laced with silvery white. The Buff has the same color pattern as the Light, with golden buff replacing the white. The Dark has silver-penciled plumage with beautiful feathering, some greenish black laced with silvery white.

Status: Watch

Dark Brahmas

Light Brahmas

Dark and Light Brahma Bantams

Buckeye

The Buckeye, an all-American breed, was developed in Warren, Ohio. Nettie Metcalf is credited with its development and is the only woman to create a breed in the American class. Wanting to create a dual-purpose breed with a rich brownish red color, like the buckeye nut, Metcalf first mixed a Buff Cochin male with a Barred Plymouth Rock female. She then mated a Black-Breasted Red Game with the Cochin/Plymouth Rocks and from their red offspring, she developed the Buckeye. She originally called the breed Pea Combed Rhode Island Reds but changed the name to Buckeye in 1902.

Large

Class: American
Variety: Buckeye (1904)
Weight:
Cock: 9 pounds
Hen: 6.5 pounds
Cockerel: 8 pounds
Pullet: 5.5 pounds

Bantam

Class: All Other Combs Clean Legged Bantams
Variety: Pea comb (1960)

Characteristics

Comb: Moderately small pea
Tail: Medium length, the male's is carried at a 40-degree angle, the female's at 30 degrees.
Legs: Yellow shanks and toes. The legs are stout, smooth, and set well apart.
Eggs: Medium, brown
Other: The plumage is an overall rich mahogany bay, and there is a slate-colored bar in the fluff of the back.

Status: Critical

Buckeyes (Courtesy of William A. Suys, Jr.)

Campine

The Campine is an old breed descended from the Turkish fowl that were mentioned by Ulisse Aldrovandi, the Renaissance chicken expert. It was developed in Belgium in the district of La Campine, principally for its large white eggs. Brought to England around 1855, Campines became popular exhibition birds. The Campine has the same basic ancestry as the Belgian-bred Braekel but is smaller. A Campine club formed in England in 1899; the first one in America formed in 1911.

Silver Campines

Large

Class: Continental (North European)
Varieties: Silver, Golden (1914)
Weight:
Cock: 6 pounds
Hen: 4 pounds
Cockerel: 5 pounds
Pullet: 3.5 pounds

Bantam

Class: Single Comb Clean Legged Other Than Game Bantams
Varieties: Silver, Golden (1960)

Characteristics

Comb: Medium single with five serrations, erect on the male's head and lopped on the female's
Tail: Long and full, the male's is carried at a 45-degree angle, the female's at 40 degrees.
Legs: Leaden blue shanks and toes
Eggs: Medium, white
Other: Both sexes have identical color patterns, and the roosters should have some degree of hen-feathering. The Golden Campine has a head and neck hackle of rich gold and a barred body-feather pattern in gold bay and greenish black. The Silver Campine has a head and neck hackle of pure white and an identical body-feather pattern in white and greenish black, which gives the appearance of stripes.

Status: Critical

Catalana

The Buff Catalana originated in the Catalonia region of Spain, specifically the district of Prat. The breed is sometimes called the Catalana del Prat Leonada or simply the Prat. It was introduced at the 1902 Madrid World Fair. Though not common in the United States, it is popular in South America and Latin America. A dual-purpose breed, the Catalana is known for its large eggs and good meat and is described as hardy and vigorous.

Buff Catalanas (Courtesy of William A. Suys, Jr.)

Large

Class: Mediterranean
Variety: Buff (1949)
Weight:
Cock: 8 pounds
Hen: 6 pounds
Cockerel: 6.5 pounds
Pullet: 5 pounds

Bantam

Class: Single Comb Clean Legged Other Than Game Bantams
Variety: Buff (1960)

Characteristics

Comb: Single. The male's is large and erect with six points plus a blade toward the back of the neck. The front portion of the female's stands erect and the rest droops to one side.
Tail: Medium-sized, black, carried at a 45-degree angle
Legs: Bluish slate shanks and toes
Eggs: Medium or large, white or tinted light brown
Other: The male's plumage is varying shades of buff, from light to reddish to dark. He sports a lustrous greenish black tail. The overall coloration of the female is buff with a black tail and two top feathers often edged with buff.

Status: Critical

Chantecler

The Chantecler was developed in Quebec, Canada, at the Oka Agricultural Institute. Breeders spent ten years developing the fowl and introduced it in 1918. The breeders' goal was to create a strong, dual-purpose bird adapted to the harsh Canadian winters, so they developed a breed with a small comb and wattles to avoid frostbite. The breeds used to develop the White variety were the Dark Cornish, Rhode Island Red, White Leghorn, White Plymouth Rock, and White Wyandotte. The Partridge variety, created in Edmonton, Alberta, was developed from the Dark Cornish, Partridge Cochin, Partridge Wyandotte, and Rose Comb Brown Leghorn.

Large

Class: American
Varieties: White (1921), Partridge (1935)
Weight:
Cock: 8.5 pounds
Hen: 6.5 pounds
Cockerel: 7.5 pounds
Pullet: 5.5 pounds

Bantam

Class: All Other Combs Clean Legged Bantams
Varieties: White, Partridge (1960)

Characteristics

Comb: Small cushion-shaped
Tail: Large and full, the male's is carried at a 45-degree angle and the female's at 30 degrees.
Legs: Yellow shanks and toes
Eggs: Large, brown
Other: Small wattles

Status: Critical

White Chanteclers (Courtesy of William A. Suys, Jr.)

Cochin

Much has been written and disputed about the origin of the Cochin. It is generally agreed that Cochins, originally called Shanghais, were first exported from the Chinese seaport town of Shanghai in the mid-1800s. Their name was later changed to Cochin China, to reflect the belief that the chicken actually originated in the historic region of Vietnam called Cochin China. Eventually, the breed's name was shortened to simply Cochin. They were introduced after the China War in 1845, when poultry enthusiast Queen Victoria received them as a gift. She showed them off in her opulent royal poultry yard and exhibited them at the Show of the Royal Dublin Agricultural Society in 1846. They were described then as gigantic, were thought to be related to buzzards, and acquired the name "ostrich fowl." Some were described as having legs long enough to step over a high fence. The feathered shanks of today's Cochins were not the preference at that time.

Cochin bantams, called Pekins in Australia and Great Britain, are popular as show birds and as pets. Like their large counterparts, Cochin bantams are friendly, broody, and wonderful mothers.

Continued p. 84...

White Cochins

Partridge Cochins

Buff, Partridge, Black, and White Cochin Bantams

Large

Class: Asiatic
Varieties: Buff, Partridge, White, Black (1874), Silver Laced, Golden Laced, Blue, Brown (1965), Barred (1982)
Weight:
Cock: 11 pounds
Hen: 8.5 pounds.
Cockerel: 9 pounds
Pullet: 7 pounds

Bantam

Class: Feather Legged Bantams
Varieties: Black, Buff, Partridge, White (1874), Barred, Brown Red, Golden Laced, Mottled, Silver Laced (1965), Birchen, Blue, Columbian, Red (1977)

Characteristics

Comb: Single with five points. The male's is medium and the middle point is the largest; the female's is small with even points.
Tail: Round, full, and short
Legs: Yellow shanks and toes. Large globes of feathering entirely conceal legs. Plumage is also on the feet.
Eggs: Small, brown
Other: An excessive amount of long feathers and an abundance of down in the under fluff give Cochins a rotund appearance.

Status: Watch

Black Cochins

Buff Cochins

Cornish

As early as 1820, the Cornish was developed in Devon and Cornwall, south west England. Bred for cockfighting, they were created by mixing Malays, Red Aseels, and Black-Breasted Red Pit Game Fowls. Sir Walter Raleigh Gilbert is said to be the originator of the breed. Later, they were developed into exhibition fowl called Indian Game. The original breed, described as "a blocky Aseel-like game bird" has evolved into a stocky show bird that, when viewed from above, is heart shaped. Admitted into the *American Standard* as the Cornish Indian Game, by 1910 they were called Cornish fowls.

Dark Cornish

Large

Class: English
Varieties: Dark (1893), White (1998), White Laced Red (1909), Buff (1938)
Weight:
Cock: 10.5 pounds
Hen: 8 pounds
Cockerel: 8.5 pounds
Pullet: 6.5 pounds

Bantam

Class: All Other Combs Clean Legged Bantams
Varieties: Dark (1933), White, White Laced Red (1942), Buff (1960), Blue Laced Red, Spangled (1965), Black, Mottled (1996)

Characteristics

Comb: Moderately small pea
Tail: Short, carried slightly below the horizontal
Legs: Shanks and toes are various shades of yellow. Legs are short, strong, and set widely apart.
Eggs: Small, light brown
Other: Squat, stock body configuration in both sexes, though the male is larger. The feathers are short, hard, and close fitting.

Status: Watch

Crevecoeur

The Crevecoeur is the oldest of the standard-bred French fowl, according to poultry expert W. B. Tegetmeier. At the first great agricultural show in Paris in 1855, there were two sets of prizes: one for the Crevecoeur and the other for all other breeds. The breed originated in the village of Crevecoeur, in the Normandy region, and was a mixture of Polish and other fowl. They were first referred to as Black Polish. Developed for their white egg production and easily fattened, they became one of the favored gourmet breeds.

Large

Class: Continental (French)
Variety: Black (1874)
Weight:
Cock: 8 pounds
Hen: 6.5 pounds
Cockerel: 7 pounds
Pullet: 5.5 pounds

Bantam

Class: All Other Combs Clean Legged Bantams
Variety: Black (1960)

Characteristics

Comb: V-shaped
Tail: Long and full, carried at a 45-degree angle
Legs: Leaden blue shanks and toes
Eggs: Medium, white
Other: A large oval crest and full beard hides the head, except for the comb, beak, and wattles. The plumage is glossy black with lustrous greenish black tones.

Status: Critical

Black Crevecoeurs

Cubalaya

The Cubalaya is thought to have its roots in Oriental game fowl brought to Cuba from the Philippine Islands, as well as in Old English and Spanish fowl. This combination of Oriental fighting stock and British and European utility breeds served two purposes: Cuban men could enjoy a good cockfight and the eggs and meat satisfied the needs of the family. Though these birds are profusely feathered, they have an abundance of gourmet-quality white meat and are good layers. They were first exhibited in the United States in 1939 at the Seventh World Poultry Congress. In the United States, Cubalayas are unfortunately rare and have few breeders.

Black-Breasted Red Cubalayas (Courtesy of William A. Suys, Jr.)

Large

Class: All Other Standard Breeds (Orientals)
Varieties: Black-Breasted Red, White, Black (1939)
Weight:
Cock: 6 pounds
Hen: 4 pounds
Cockerel: 4.5 pounds
Pullet: 3 pounds

Bantam

Class: All Other Combs Clean Legged Bantams
Varieties: Black, Black-Breasted Red, White (1960)

Characteristics

Comb: Pea
Tail: Long and often dragging on the ground, known as a "lobster tail" because of its downward curve
Legs: The Black-Breasted Red and White varieties have pinkish white shanks and toes; the Black's are slate. Rather than pointed, spurs are short and dome-shaped.
Eggs: Small, white
Other: The coloration of the Black-Breasted Red is notable. The male's is flamboyant, ranging from reddish chestnut to greenish black to reddish bay. The female's color is more monochromatic, ranging from reddish chestnut to light cinnamon, and lacking the black breast.

Status: Study

Delaware

The Delaware was originally called "Indian Rivers," but the name was later changed to honor the state where the breed originated. The breed was developed in 1940 by George Ellis, who crossed a Barred Plymouth Rock male and a New Hampshire female to increase the amount of meat. Until the late 1950s, the Delaware and the Delaware/New Hampshire cross were favored broiler chickens. Then the Cornish and White Rock cross began to dominate the broiler industry. Today, the Delaware is a popular dual-purpose breed.

Large

Class: American
Variety: Delaware (1952)
Weight:
Cock: 8.5 pounds
Hen: 6.5 pounds
Cockerel: 7.5 pounds
Pullet: 5.5 pounds

Bantam

Class: Single Comb Clean Legged Other Than Game Bantams
Variety: Single Comb (1960)

Characteristics

Comb: Single with five distinct points
Tail: Medium length, the male's is carried at a 40-degree angle, the hen's at 30 degrees.
Legs: Yellow shanks and toes
Eggs: Large, brown

Other: Delawares may be sex-linked. When a Delaware female is mated with a New Hampshire or Rhode Island Red male, the male offspring will have the hen's Delaware color pattern and the females will have the red coloration of the rooster.

Status: Critical

Delawares (Courtesy of William A. Suys, Jr.)

Dominique

Little is known about the origin of the Dominique, or Dominiker, although it is thought to be the oldest American breed, brought over to North America by the Puritans. In the early 1800s, poultry expert C. N. Bement praised Dominiques as "one of the best and most profitable fowls, being hardy, good layers, careful nurses, and affording excellent eggs and first quality of flesh." Due to their barred feathers, Dominiques were often confused with Barred Plymouth Rocks, although their distinctive rose comb is easily differentiated from the Barred Rocks' classic single comb. Dominiques fell out of favor in the mid-1800s after the importation of exotic Asian breeds. The decline was so dramatic that, until recently, they were nearly extinct and only shown in exhibitions. With the help of the American Livestock Breeds Conservancy and fanciers, the breed is making a comeback.

Dominiques

Large

Class: American
Variety: Dominique (1874)
Weight:
Cock: 7 pounds
Hen: 5 pounds
Cockerel: 6 pounds
Pullet: 4 pounds

Bantam

Class: Rose Comb Clean Legged Bantams
Variety: Rose Comb (1960)

Characteristics

Comb: Rose with a spike that turns slightly upward
Tail: Long and full, carried at a 45-degree angle
Legs: Yellow shanks and toes
Eggs: Medium, brown
Other: The plumage is cuckoo, having irregular dark and light bars and a dark tip on each feather.

Status: Watch

Dorking

The Dorking is one of the oldest breeds of poultry. It originated in Rome and was brought to Great Britain in the first century BC, during the Roman invasion led by Julius Caesar. Until that time, Britons did not eat fowl. A century later, the great Roman agriculturalist Columella wrote about the Dorking in his treatise "Of Husbandry in Twelve Books."

In England, the breed was developed for meat production by the farmers of Sussex and Surrey counties, and breeding records date back to 1763. In the nineteenth century, Queen Victoria favored the tasty Dorking's white breast meat, and it soared in popularity throughout England. Today, the town of Dorking, Surrey, pictures its namesake bird on its coat of arms.

Silver-Gray Dorkings

Chickens have four toes, except Dorkings, Faverolles, Houdans, Sultans, and Silkies breeds, which have five toes.

Large

Class: English
Varieties: White, Silver-Gray, Colored (1874), Red (1995), Single Comb Cuckoo, Rose Comb Cuckoo (2000)
Weight:
Cock: 9 pounds
Hen: 7 pounds
Cockerel: 8 pounds
Pullet: 6 pounds

Bantam

Bantam Dorkings are in two classes.
Class: Single Comb Clean Legged Other Than Game Bantams
Varieties: Colored (1960), Silver-Gray (1960)
Class: Rose Comb Clean Legged Bantams
Variety: Rose Comb White (1960)

Characteristics

Combs: Large single or rose. The Silver-Gray and Colored varieties have an upright single with six well-defined points, the four middle larger. The White variety has a rose with the spike turning upward.
Tail: Large and full, the male's is carried at a 45-degree angle, the female's at 35 degrees.
Legs: Pinkish white shanks and toes, relatively short legs
Eggs: Medium, creamy white to tinted light brown
Other: Dorkings are the original breed to have a fifth toe. The birds are square-framed and stout.

Status: Critical

Rose Comb White Dorkings

Dutch Bantam

When miniature chickens were developed on Indonesia's Bantam Island, they were called bantams, and the word *bantam* soon applied to any small chicken. Those that arrived in the Netherlands in the seventeenth century, brought by seamen in the East India Company during the Dutch colonization of Indonesia, were called Dutch Bantams.

The Dutch Poultry Club formed in 1946 and standardized the breed. Today, the Dutch standard recognizes over twenty varieties of the breed. Dutch Bantams continue to have a huge popularity in the Netherlands, as well as in England.

Although Dutch Bantams were first introduced to the United States in the 1940s, they did not gain popularity and became scarce. In the late 1960s, they were again imported into this country and an interest in them was sparked. The American Dutch Bantam Society formed in 1986. Today, there are five varieties in the *Standard*. The American Bantam Association, however, recognizes many other varieties, including White, Golden, Cuckoo, Crele, Blue, Self Blue, Blue Golden, Blue Silver, Blue Splash, Blue Silver Splash, Blue Light Brown Splash, Cream Blue Light Brown, and Cream Blue Light Brown Splash.

Silver Dutch Bantams

Bantam

Class: Single Comb Clean Legged Other Than Game Bantams

Varieties: Light Brown, Silver (1992), Blue Light Brown (1995), Black, Cream Light Brown (2000)

Weight:
Cock: 21 ounces
Hen: 19 ounces
Cockerel: 20 ounces
Pullet: 18 ounces

Characteristics

Comb: Medium single with five distinct points

Tail: The male's is large and carried at a 50-degree angle; the female's is smaller and carried at a 45-degree angle. Both carry their tail far back.

Legs: Slate blue shanks and toes

Eggs: Tiny, tinted light brown

Faverolles

The Faverolles was developed in the northern part of France around the communities of Faverolles and Houdan in the 1860s. The birds were created for improved egg production and meat. Polish, Crevecoeurs, Light Brahmas, Dorkings, and Houdans can be found in their bloodline. From the Dorking and the Houdan, Faverolles inherited their white flesh and legs and their fifth toe. From the Crevecoeur and the Houdan, they inherited their muff and beard, and the Brahma contributed to their leg feathering and dark-shelled eggs. The Salmon Faverolles was the first variety to be developed and is known as the fowl with the *tête de hibou*, meaning "the head of an owl." The Faverolles' popularity increased in the 1890s because it adapted to being housed in battery-type cages. The breed was standardized in France in 1893, and the Houdan Faverolles Club of France was created in 1909.

Salmon Faverolles

Large

Class: Continental (French)
Varieties: Salmon (1914), White (1981)
Weight:
Cock: 8 pounds
Hen: 6.5 pounds
Cockerel: 7 pounds
Pullet: 5.5 pounds

Bantam

Class: Feather Legged Bantams
Varieties: Salmon (1960), White (1981)

Characteristics

Comb: Medium single with five well-defined points, the three middle larger
Tail: Full, carried at a 50-degree angle
Legs: Pinkish white shanks and toes and slight leg feathering. They also have five toes.
Eggs: Medium, light brown or creamy white
Other: Faverolles have a large beard and a full muff, which are unusual with a single comb. The Salmon rooster has plumage of many colors: black beard and muff; straw hackles; a mixture of black, white, and brown on the back and shoulders; and black on the breast, body, tail, and shank. The Salmon hen is a mixture of cream and salmon.

Status: Critical

Frizzle

Frizzled chickens are thought to have originated in Java and are prevalent in the Far East, particularly Japan and the Philippines. Today, the fowl are popular throughout the world. Charles Darwin called the birds Frizzled or Caffie Fowls. Frizzles get their name from the strange nature of their feathers. Each feather curls, or frizzles, toward the head. Breeder T. Farrar Rackham of New Jersey was instrumental in their development in the United States. Though the Frizzle is listed in the *American Standard* as a breed, it is not actually a separate breed with a set comb style, weight, or tail carriage. The genetic variation of frizzled feathers can be introduced into many breeds, and any frizzled chicken can be exhibited in the Frizzle breed category.

Clean Legged Frizzles

Large

Class: All Other Standard Breeds (Miscellaneous)
Varieties: Clean Legged (1874), Feather Legged (1982)

Bantam

Class: Feather Legged Bantams
Variety: Feather Legged (1951)

Hamburg

The Hamburg is an old breed, identified by the Italian naturalist Ulisse Aldrovandi in 1599 as a Turkish chicken. He describes the birds as having white or yellow bodies, feathers spangled with black spots, and feet tinged with blue. From Italy, the breed was exported to the Netherlands and took on the name Hollands Hoen, or Dutch Fowl. The Netherlands lays claim to the development of the Silver and Golden Penciled varieties. The Germans created the Black variety, using the Sumatra and the Minorca. Eventually, the breed earned the name Hamburg from the German port city through which many of the chickens were exported. The Gold and Silver Spangled varieties are thought to have originated in Great Britain. Although a good egg layer, once called the "everyday layer," the Hamburg declined in popularity in England when the Leghorn, a more prolific producer of larger eggs, was introduced in 1982.

Continued p. 96...

Golden Penciled Hamburgs

Large

Class: Continental (North European)
Varieties: Gold Spangled, Silver Spangled, Golden Penciled, Silver Penciled, White, Black (1874)
Weight:
Cock: 5 pounds
Hen: 4 pounds
Cockerel: 4 pounds
Pullet: 3.5 pounds

Bantam

Class: Rose Comb Clean Legged Bantams
Varieties: Silver Spangled (1939), Black, Golden Penciled, Golden Spangled, Silver Penciled, Silver Spangled, White (1960)

Characteristics

Comb: Medium rose with a spike that curves upward
Tail: Moderately long, the male's is carried at a 40-degree angle, the female's at 35 degrees.
Legs: Leaden blue shanks and toes, pinkish white on the bottoms of the feet
Eggs: Small, white
Other: The varieties vary in size with the Black and White the largest and the Penciled the smallest.

Status: Watch

Silver Spangled Hamburgs

Silver Penciled Hamburgs

Holland

The Holland is an American breed created in 1934 from imported Dutch fowl crossed with White Leghorns, Rhode Island Reds, New Hampshires, and Lamonas. The result was the White Holland. The Barred Holland was developed from a mixture of White Leghorns, Barred Plymouth Rocks, Australorps, and Brown Leghorns. The goal was to create a white-egg laying, dual-purpose chicken.

Large

Class: American
Varieties: White, Barred (1949)
Weight:
Cock: 8.5 pounds
Hen: 6.5 pounds
Cockerel: 7.5 pounds
Pullet: 5.5 pounds

Bantam

Class: Single Comb Clean Legged Other Than Game Bantams
Varieties: White, Barred (1960)

Characteristics

Comb: Moderately large single with six well-defined points, the four middle larger
Tail: Moderately long, the male's is carried at a 45-degree angle, the female's at 40 degrees.
Legs: Yellow shanks and toes
Eggs: Medium, white
Other: The Holland is one of two breeds in the *Standard* that have red earlobes and lay white eggs. (The Lamona is the other.)

Status: Critical

White Hollands (Courtesy of William A. Suys, Jr.)

Houdan

The Houdan, recognized as one of the oldest French breeds, originated in the northern French town of Houdan. The bird existed during the reign of Louis XIII, in the seventeenth century, but was developed in the mid-1800s. It was called the Dorking of France because of its fifth toe. The Houdan's large headdress and beard give it a distinctive appearance. Said to be from intermingling Black Polish, Crevecoeur, and Dorking, the breed was introduced into England in 1850 and bred there for exhibition. The first American record of the Houdan is from 1859 in Philadelphia.

Mottled Houdans

Large

Class: Continental (French)
Varieties: White (1874), Mottled (1914)
Weight:
Cock: 8 pounds
Hen: 6.5 pounds
Cockerel: 7 pounds
Pullet: 5.5 pounds

Bantam

There is no Houdan Bantam in the *American Standard*.

Characteristics

Comb: V-shaped. Although early Houdans are described as having a leaf comb, a change in the *Standard* in 1883 declared it to be a horn comb shaped like the letter V.
Tail: Full, the male's is carried at an angle of 50 degrees and the female's at 35 degrees.
Legs: The White variety has pinkish white shanks and toes; the Mottled variety has pinkish white with black mottling. They have five toes.
Eggs: Small to medium, white
Other: A flamboyant crest and full beard nearly hide the face. The Mottled variety has beautiful feathers that are greenish black with V-shaped white tips.

Status: Critical

Japanese Bantam

The Japanese Bantam, also called the Chabo, was created in Japan from chickens imported from South China. This prized bird was guarded in the emperor's royal gardens as far back as the Tang Dynasty (AD 618–906). Beginning in the thirteenth century, emperors permitted the movement of these birds throughout Japan. The early Japanese Bantams were called Creepers, because their short legs made the birds appear as though they were sliding along on their wings, which seemed to serve as runners.

The Chabo arrived in the Netherlands in the seventeenth century, presumably via the same East India Company trading ships that brought the Dutch Bantam from Indonesia. The Dutch artist Jan Steen depicts Japanese Bantams in his 1660 painting *De Hoenderhof*, or *The Chicken Run*. They arrived in England around 1860 and were exhibited there in 1910. A club, originally formed in 1921, took a break during World War II and regrouped in 1961. The *Poultry Standard of Japan* admitted the breed in 1941.

Black-Tailed White Japanese Bantams

Bantam

Class: Single Comb Clean Legged Other Than Game Bantams

Varieties: Black-Tailed White (1874), Black, White (1883), Gray (1914), Mottled (1947), Black-Tailed Buff (1982), Brown Red, Wheaten (1996)

Weight:
Cock: 26 ounces
Hen: 22 ounces
Cockerel: 22 ounces
Pullet: 20 ounces

Characteristics

Comb: Single. The male's is a large with five distinct points. The female's is significantly smaller.

Tail: Long, erect, carried well forward, almost touching the back of the head

Legs: Yellow shanks and toes; short, smooth legs.

Eggs: Tiny, creamy white

Other: Profusely feathered with long hackles

Java

Black Javas (Courtesy of William A. Suys, Jr.)

The Java is one of the oldest breeds in America, known to be in the United States as early as 1835. Though the fowl got its name from the Indonesian island of Java, where it is thought to have originated, much modification occurred in the United States. Daniel Webster, an early poultry enthusiast, exhibited a pair of Java fowls at the first poultry show in Boston in 1849. The Black and Mottled varieties reached their peak popularity for their meat production in the mid-1800s. Their popularity fell when they were crossed with other breeds to produce the Jersey Giant and Plymouth Rock.

Chicago's Museum of Science and Industry became involved in reviving the nearly extinct Black Javas and, in 1999, hatched many chickens as children and adults watched. Amazingly, through recessive gene reproduction, two of the seventy hatched eggs were White Javas, unseen for the past fifty years. Now, thanks to the Museum of Science and Industry and the Garfield Farm Museum in Illinois, which works to preserve rare breed livestock, the Black, White, and even a Blue variety are making a comeback.

Large

Class: American
Varieties: Black, Mottled (1883)
Weight:
Cock: 9.5 pounds
Hen: 7.5 pounds
Cockerel: 8 pounds
Pullet: 6.5 pounds

Bantam

Class: Single Comb Clean Legged Other Than Game Bantams
Varieties: Black, Mottled (1960)

Characteristics

Comb: Moderately small single. The first point should be above the eye instead of the nostril.
Tail: Long and full, the male's is carried at an angle of 55 degrees, the female's at 45 degrees.
Legs: The Black variety's shanks and toes are black and the bottoms of its feet yellow. The Mottled's shanks and toes are leaden blue with yellow bottoms.
Eggs: Medium, brown
Other: The Black Java is noted for its brilliant beetle green coloration and nearly black eyes. The Mottled Java has red eyes.

Status: Critical

Jersey Giant

Black Jersey Giants

The Jersey Giant lives up to its name, being the largest American breed. It was developed in New Jersey during the 1870s and '80s by brothers with the last name Black, who called it Jersey Black Giants. Breeds used to create the Jersey Giant were the Black Java, Black Langshan, Dark Brahma, and Orpington. The Black brothers were aiming for a large fowl that could be used for capon production. Unfortunately, the Jersey Giant is a slow grower that develops a large frame first and has little meat at six months. The breed's popularity continues because of the birds' size.

Large

Class: American
Varieties: Black (1922), White (1947), Blue (2000)
Weight:
Cock: 13 pounds
Hen: 10 pounds
Cockerel: 11 pounds
Pullet: 8 pounds

Bantam

Class: Single Comb Clean Legged Other Than Game Bantams
Varieties: Black, White (1960)

Comb: Moderate single with six well-defined and evenly serrated points

Tail: Rather large, the male's is carried at 45 degrees, the female's at 30 degrees.
Legs: Dark willow shanks and toes
Eggs: Medium, brown to dark brown

Status: Watch

La Flèche

The La Flèche is an old French breed that originated in the small town of La Flèche in northwest France near Le Mans. The word *flèche* means "arrow" in French and describes the shape of the bird's comb. According to French experts, the La Flèche has Black Spanish, Crevecoeur, Black Polish, Minorca, and other French breeds in its bloodline. Because of the flavor and abundance of its white breast meat, it was popular in the Paris markets in the mid-1800s. In 1859, the La Flèche was introduced to Germany, where a club formed and the breed thrived until World War II. By the end of the war, the La Flèche had disappeared. Similarly, in France, a La Flèche club formed in 1905, then the breed declined in popularity. But the French club was reactivated in 1985, and today the breed is making a comeback. The La Flèche was admitted to the first *American Standard* in 1874.

Large

Class: Continental (French)
Variety: Black (1874)
Weight:
Cock: 8 pounds
Hen: 6.5 pounds
Cockerel: 7 pounds
Pullet: 5.5 pounds

Bantam

Class: All Other Combs Clean Legged Bantams
Variety: Black (1960)

Characteristics

Comb: Rather large V-shaped
Tail: Very long, the male's is carried at an angle of 45 degrees, the female's at 40 degrees.
Legs: Dark slate shanks and toes
Eggs: Large, white
Other: The plumage is a lustrous greenish black. The breed has distinctive white earlobes and long dangling wattles.

Status: Critical

Black La Flèche

Lakenvelder

The Lakenvelder originated in Holland, bred in the same locality as the Campine, but Germany lays claim to the breed's development. The name may come from the Dutch word for "shadow on a sheet" or possibly is derived from the hamlet of Lakenveld, near Utrecht in Holland. Lakenvelders are mentioned in a travel story from 1727. There is record of them in 1835 in West Hanover, where they were shown. In 1902, they were exhibited in England and, around the same time, made their way to America.

Large

Class: Continental (North European)
Variety: Lakenvelder (1939)
Weight:
Cock: 5 pounds
Hen: 4 pounds
Cockerel: 4 pounds
Pullet: 3.5 pounds

Bantam

Class: Single Comb Clean Legged Other Than Game Bantams
Variety: Single Comb (1960)

Characteristics

Comb: Medium single, erect with five points, not too deeply serrated
Tail: Long and broad main tail feathers and sickles. The male's is carried at a 45-degree angle, the female's at 40 degrees.
Legs: Slate-colored shanks and toes
Eggs: Small, white or tinted light brown
Other: The head, neck, and tail are a rich black. The rest of the body is white on the surface.

Status: Rare

Lakenvelders (Courtesy of William A. Suys, Jr.)

Lamona

It is speculated that the Lamona, an American creation, is near extinction. Currently, there are only a few breeders attempting to revive the breed. The Lamona was developed by Harry Lamon, a New Yorker with a long list of poultry credentials: breeder of top prize-winning Buff Leghorns at the Madison Square Garden Poultry Show; reputed to be one of the best poultry judges in the world; founder of the National Poultry Institute; editor of the *National Poultry Journal*; and president of the International Association of Poultry Husbandry. Hired by the U.S. government in 1910, he was soon promoted to Senior Poultryman for the U.S. Department of Agriculture's experimental station in Beltsville, Maryland. It was here that he conceived and established the breed named in his honor. Using Silver Gray Dorkings, White Plymouth Rocks, and White Leghorns, he spent sixteen years developing the breed to mature quickly, produce large white eggs, and have fine and abundant meat even after egg production waned.

Lamonas (Courtesy of William A. Suys, Jr.)

Large

Class: American
Variety: Lamona (1933)
Weight:
Cock: 8 pounds
Hen: 6.5 pounds
Cockerel: 7 pounds
Pullet: 5.5 pounds

Bantam

Class: Single Comb Clean Legged Other Than Game Bantams
Variety: White (1960)

Characteristics

Comb: Moderately large single, straight with five evenly serrated points, the middle three larger
Tail: Large and full. The male's is carried at a 45-degree angle, the female's at 40 degrees.
Legs: Rich yellow shanks and toes
Eggs: Medium to large, white
Other: The Lamona is one of two breeds in the *Standard* that has red earlobes and lays white eggs. (The Holland is the other.)

Status: Study, may be extinct

Langshan

When a British military officer named Croad returned to Sussex from the Langshan district of northern China in 1872 with some black chickens, they became known as Croad Langshans. They looked similar to Black Cochins of that period and became popular because of their dark brown eggs. A Croad Langshans club formed in England in 1904. Today's Langshans, an offshoot of the Croad Langshans, are sometimes referred to as Modern Langshans. Their legs are longer and less feathered than the Croad's, and they serve primarily as exhibition fowl.

Black Langshans

Large

Class: Asiatic
Varieties: Black (1883), White (1893), Blue (1987)
Weight:
Cock: 9.5 pounds
Hen: 7.5 pounds
Cockerel: 8 pounds
Pullet: 6.5 pounds

Bantam

Class: Feather Legged Bantams
Varieties: Black, White (1960), Blue (1987)

Characteristics

Comb: Medium single, upright with five evenly serrated points
Tail: Large, the male's is carried at an angle of 75 degrees, the female's at 70 degrees.
Legs: Dark shanks and toes. The legs are feathered on the outside and on the outer toes; feathering matches body coloration.
Eggs: Medium, dark brown

Status: Rare

Leghorn

The Leghorn district of Italy was the home of the original Leghorn fowls. They were brought to the United States from 1835 to 1853 on Spanish sailing vessels whose captains sold their surplus stock upon arrival. The breed arrived in England from America around 1870. For about a hundred years, the Leghorn was considered indispensable in the poultry industry because of its prolific egg laying as well as its meat production. The bird then played a prominent role in the chicken industry as its hybrids became egg-producing machines. Leghorns come in many colors and their combs vary, but they all have a similar body, size, and tail shape. There are sixteen varieties, all admitted to the *American Standard* at various times. The American Brown Leghorn club, founded in 1910, specializes in the Dark and Light Brown varieties of large and bantam Leghorns.

Large

Class: Mediterranean
Varieties: Single Comb Dark Brown, Single Comb Light Brown, Single Comb White, Single Comb Black (1874), Rose Comb Dark Brown, Rose Comb Light Brown, (1883), Rose Comb White (1886), Single Comb Buff, Single Comb Silver (1894), Silver (1910), Single Comb Light Brown, Single Comb Dark Brown (1923), Single Comb Red, Single Comb Black-Tailed Red, Single Comb Columbian, (1929), Rose Comb Light Brown, Rose Comb Dark Brown (1933), Rose Comb Black, Single Comb Buff, Rose Comb Silver, Single Comb Golden Duckwing (1981)
Weight:
Cock: 6 pounds
Hen: 4.5 pounds
Cockerel: 5 pounds
Pullet: 4 pounds

Bantam

Leghorn bantams are in two classes.
Class: Single Comb Clean Legged Other Than Game Bantams
Varieties: Black, Black-Tailed Red, Buff, Columbian, Dark Brown, Light Brown, Red, Silver (1960), White (1940), Golden Duckwing (1981), Barred (1995)
Class: Rose Comb Clean Legged Bantams
Varieties: Black, Buff, Silver (1981), Dark Brown, Light Brown, White (1960)

Characteristics

Comb: Medium single or medium rose. The male single has five deeply serrated points that extend well over the back of the head on the male; the female single has a similar though smaller configuration, and the back four points droop gradually to one side. The rose is square in front and tapers to a defined spike that extends horizontally well over the back of the head.
Tail: Large, the male's is carried at an angle of 40 degrees, the female's at 35 degrees.
Legs: Yellow shanks and toes
Eggs: Medium, white

Single Comb White Leghorns

Malay

The Malay, said to have descended from the Great Malay or Kulm fowls of India, is one of the oldest breeds of poultry. Although similar to the Aseel, it did not have the same fighting instinct. Instead, it was popular for its meat production. Malays came to England from Malaysia around 1830 and were developed in the Cornwall and Devon areas as Malay Game Fowls. They were shown at the first poultry show in England in 1845, and both the Black-Red and White Malays appear in the first *British Book of Standards* of 1865. Around 1900, the Malay was the first large breed chicken to be bantamized; the new miniature became very popular.

Large

Class: All Other Standard Breeds (Orientals)
Varieties: Black Breasted Red (1883), Black, Red Pyle, Spangled, White (1981), Wheaten female (1996)
Weight:
Cock: 9 pounds
Hen: 7 pounds
Cockerel: 7 pounds
Pullet: 5 pounds

Bantam

Class: All Other Combs Clean Legged Bantams
Varieties: Black Breasted Red (1904), Black, Red Pyle, Spangled, White (1981), Wheaten female (1996), Wheaten male (2000)

Characteristics

Comb: Moderately small strawberry, close to the beak
Tail: Medium sized, droops below the horizontal. Feathers are well folded together.
Legs: Shanks and toes in varying shades of yellow
Eggs: Medium, tinted brown
Other: The Malay stands upright and has a firm and muscular carriage. Its distinctive face has an overhanging skull and large sunken eyes that give it a cruel expression. The face and throat have no feathers.

Status: Critical

Wheaten, Black Breasted Red, Red Pyle, and Spangled Malays and Malay Bantams

Minorca

The Minorca originated in the Mediterranean and was named after the Spanish island of Minorca. The breed was imported to southwest England in the early 1830s and developed for its extra-large white eggs. It is sometimes referred to as the Spanish Red Face. Though larger, the Minorca is similar to the White Faced Black Spanish but has a red face instead of white.

Single Comb Black Minorcas

Large

Class: Mediterranean
Varieties: Single Comb Black, Single Comb White (1888), Rose Comb Black (1904), Single Comb Buff (1913), Rose Comb White (1914)
Weight:
Cock: 9 pounds
Hen: 7.5 pounds
Cockerel: 7.5 pounds
Pullet: 6.5 pounds

Bantam

Minorca bantams are in two classes.

Class: Single Comb Clean Legged Other Than Game Bantams
Varieties: Black (1944), Buff, White (1960)
Class: Rose Comb Clean Legged Bantams
Varieties: Black, White (1960)

Characteristics

Comb: Large single or moderately large rose. The single has six distinct points. On the male, the middle point is the longest and the same length as the width of the blade. The female's single forms a loop over the beak and droops down the opposite side of the head. The rose is square in front and tapers to a spike that follows the contour of the head.
Tail: Long, carried at a 35-degree angle
Legs: The Black variety has dark slate shanks and toes; the White and Buff have pinkish white.
Eggs: Large, white
Other: The Minorca has large white earlobes, long wattles, and a long back.

Status: Watch

Rose Comb White Minorcas

Modern Game

Today, the *American Standard* recognizes two breeds of large game chickens: Modern Games and Old English Games. The Modern evolved from the Old English over the years, and their differences in appearance are striking. At the first poultry show in England in 1845, the Old English represented the only game fowl. By 1850, the Old English body formation had not changed, but its coloration was more pronounced and its comb and wattle more trimmed. For the next twenty-five years, the Old English was altered with great care into the Modern Game, a separate breed that stands upright with a tall and slender appearance, long muscular shanks and thighs, a long, slightly arched neck, and a "slim tail" on the rooster. The shortness and firmness of the feathers was also important, and in show, the comb, wattles, and earlobes were dubbed.

Large

Class: All Other Standard Breeds (Games)
Varieties: Black Breasted Red, Brown Red, Golden Duckwing, Silver Duckwing, Birchen, Red Pyle, White, Black (1874), Wheaten (1981)
Weight:
Cock: 6 pounds
Hen: 4.5 pounds
Cockerel: 5 pounds
Pullet: 4 pounds

Bantam

Class: Game Bantams
Varieties: Black, Black Breasted Red, Brown Red, Golden Duckwing, Red Pyle, Silver Duckwing, White (1874), Blue, Blue Breasted Red, Lemon Blue (1965), Wheaten (1981), Silver Blue (1996)

Characteristics

Comb: Single. The male's is dubbed for exhibition, and the female's is small with five evenly serrated points.
Tail: Short and compact, carried almost horizontally
Legs: Long, muscular lower thighs and long, round shanks. Shank and toe color varies within the **varieties:** All are dark—with shades from black, brown, to willow—except the Red Pyle, White, and Wheaten, which have yellow shanks and toes.
Eggs: Small, white to tinted light brown

Status: Study

Black Breasted Red Modern Games

Naked Neck

The Naked Neck has a lot of monikers, including the Transylvanian Naked Neck, Turken, Old Welsh, and Churkey. It is called the Kaalneck in South Africa, where the breed is particularly popular. It is thought to have originated in Romania or Hungary and been developed in Austria and Germany. The Naked Neck is aptly named, because it has a featherless area between its shoulders and the base of its skull. It has less than half the number of feathers than other fowl of similar size, which makes dressing it easier than dressing other fowl.

Large

Class: All Other Standard Breeds (Miscellaneous)
Varieties: Red, White, Buff, Black (1965)
Weight:
Cock 8.5 pounds
Hen 6.5 pounds
Cockerel 7.5 pounds
Pullet 5.5 pounds

Bantam

Class: Single Comb Clean Legged Other Than Game Bantams
Varieties: Black, Red, White (1965)

Characteristics

Comb: Medium single with five evenly serrated points
Tail: Medium length, carried at a 20-degree angle
Legs: The Red, White, and Buff varieties have yellow shanks and toes; the Black's are black.
Eggs: Medium, creamy light brown
Other: Featherless neck

Naked Necks

New Hampshire

Around 1915, breeders in the Granite State began developing the New Hampshire, often called the New Hampshire Red. They used Rhode Island Reds as their base to develop this strong, fast-maturing chicken that lays brown eggs. The breed was standardized in 1935.

New Hampshires

Large

Class: American
Variety: New Hampshire (1935)
Weight:
Cock: 8 pounds
Hen: 6.5 pounds
Cockerel: 7.5 pounds
Pullet: 5.5 pounds

Bantam

Class: Single Comb Clean Legged Other Than Game Bantams
Variety: Single Comb (1960)

Characteristics

Comb: Moderately large single with five distinct points
Tail: Medium length, the male's is carried at a 45-degree angle, the female's at 35 degrees.
Legs: Yellow shanks and toes with a reddish tinge. In the male, a red line runs down the sides of the shanks to the tip of the toes.
Eggs: Large, light to medium-dark brown

Status: Watch

Old English Game

Game Fowls are said to be the direct ancestors of the red jungle fowl. Capable of defending themselves in the wild and equipped with a pugnacious nature, game fowls were naturals at cockfighting. Originally called Pit Game Fowls, they were bred for exhibition after cockfighting was outlawed in England in 1849 and became known as Old English Game. There are twenty-four varieties of Old English Game Bantams in the *American Standard*.

Large

Class: All Other Standard Breeds (Games)
Varieties: Black Breasted Red, Brown Red, Golden Duckwing, Silver Duckwing, Red Pyle, White, Black, Spangled (1938), Blue Breasted Red, Lemon Blue, Blue Golden Duckwing, Blue Silver Duckwing, Self Blue (1965), Crele (1996), Fawn Silver Duckwing (2000)
Weight:
Cock: 5 pounds
Hen: 4 pounds
Cockerel: 4 pounds
Pullet: 3.5 pounds

Bantam

Class: Game Bantams
Varieties: Black Breasted Red, Spangled (1925), Black, Golden Duckwing, Silver Duckwing, White (1938), Wheaten female (1943), Red Pyle (1946), Wheaten male (1949), Brown Red (1960), Blue Breasted Red, Blue Golden Duckwing, Blue Silver Duckwing, Lemon Blue, Self Blue (1965), Birchen, Blue, Crele (1976), Cuckoo (1977), Ginger Red (1982), Quail (1988), Brassy Back (1990), Blue Brassy Back, Columbian, Silver Blue (1996), Fawn Silver Duckwing (1998), Mille Fleur (2000)

Characteristics

Comb: Single. The male's is dubbed for exhibition; the female's is small with five points.
Tail: Large, carried at a 45-degree angle
Legs: Black, white, pinkish white, or dark blue shanks and toes, depending on variety
Eggs: Small, white or tinted light brown

Status: Study

Golden Duckwing and Black Breasted Red Old English Games

Orpington

The Black Orpington was first developed by William Cook in Orpington, in the county of Kent, England. Cook wanted a fowl with good egg production and meat quality, as well as one that would be attractive for exhibition. He bred for black plumage, which would not show the dirt and soot of London. The first to interbreed Mediterranean, Asiatic, and American breeds, Cook crossed Black Minorca males with American Barred Plymouth Rock females. He then mated the black offspring pullets with Black Langshan males. He introduced the Black variety of Orpington in 1886, the White in 1889, and the Buff in 1894, according to the *British Poultry Standards*. The breed arrived in the United States around 1890, where, after being shown in Madison Square Garden in 1895, its popularity soared. The breed has changed throughout the years and although many varieties have existed, the *American Standard* recognizes only the Buff, Black, White, and Blue.

Buff Orpingtons

Black Orpingtons

Large

Class: English
Varieties: Buff (1902), Black, White (1905), Blue (1923)
Weight:
Cock: 10 pounds
Hen: 8 pounds
Cockerel: 8.5 pounds
Pullet: 7 pounds

Bantam

Class: Single Comb Clean Legged Other Than Game Bantams
Varieties: Black, Blue, Buff, and White (1960)

Characteristics

Comb: Medium single, upright with five distinct points, the middle larger
Tail: Moderately long, the male's is carried at an angle of 25 degrees and the female's at 15 degrees.
Legs: Moderately short and stout. The Buff and White varieties have pinkish white legs, the Black has dark slate, and the Blue has leaden blue. The bottoms of the feet are pinkish white in all varieties.
Eggs: Medium to large, light to dark brown

Status: Recovering

Phoenix

Originally from Japan, the Phoenix was developed in Europe and the United States from the long-tailed Onagadori. The length of the Phoenix's tail, however, does not reach that of the Onagadori's record length of forty feet! Although the Phoenix once sported a tail of twenty feet, today's tails are more in the three-foot range and spread out like a fan.

Golden Phoenix (Courtesy of William A. Suys, Jr.)

Large

Class: All Other Standard Breeds (Oriental)
Varieties: Silver (1965), Golden (1983)
Weight:
Cock: 5.5 pounds
Hen: 4 pounds
Cockerel: 4.5 pounds
Pullet: 3.5 pounds

Bantam

Class: Single Comb Clean Legged Other Than Game Bantams
Varieties: Silver (1965), Golden (1983)

Characteristics

Comb: Single. The male's is moderate with five distinct points. The female's is similar but much smaller.
Tail: Long and large, the male's is carried at a 25-degree angle, the female's at 20 degrees.
Legs: Light leaden blue shanks and toes
Eggs: Small, white to tinted brown
Other: The Silver male and female have different plumage. The male's coloration is basically a lustrous greenish black and silvery white. There is a distinctive duckwing-like bar on the wing. The female has various shades of salmon on the neck and breast, with feathers ranging from black to various shades of gray and white elsewhere. The Golden Phoenix has the same plumage as the Golden Duckwing Leghorn.

Status: Study

Plymouth Rock

The first Plymouth Rocks were shown in 1849 in Boston at the first poultry show in America. This bird was said to be a cross of the Cochin, Dorking, Malay, and red jungle fowl and had many variations of plumage and leg color. This variety is thought to have disappeared, and the present-day Plymouth Rock is a re-creation of the original breed. Though several people were involved in the development of today's Barred Plymouth Rock, the variety is often credited to the Reverend D. A. Upham of Worcester, Massachusetts. He crossed a Black Java female with a Dominique male and mated the offspring with Dominiques. This variety made its debut at the Worcester poultry show in 1869. Others claim that the Barred Rock came from crosses of Dominique, Java, Cochin, and perhaps Malay and Dorking. By 1882, according to *The Complete Poultry Book*, the Plymouth Rock was the most popular breed in America, not only because of its excellent egg and meat production but also because of its hardiness and docility.

Buff Plymouth Rocks

Barred Plymouth Rocks

Large

Class: American

Varieties: Barred (1874), White (1888), Buff (1894), Silver Penciled (1907), Partridge (1909), Columbian (1910), Blue (1920)

Weight:

Cock: 9.5 pounds

Hen: 7.5 pounds

Cockerel: 8 pounds

Pullet: 6 pounds

Bantam

Class: Single Comb Clean Legged Other Than Game Bantams

Varieties: Barred (1940), White (1944), Blue, Buff, Columbian, Partridge, Silver Penciled (1960), Black (1991)

Characteristics

Comb: Medium single with five evenly serrated points, the middle three larger

Tail: Medium length, the male's is carried at a 30-degree angle, the female's at 20 degrees.

Legs: Yellow shanks and toes

Eggs: Large, light or pinkish to medium brown

Other: The Barred variety has exquisite plumage, each feather having well-defined horizontal bars of black and white.

Status: Recovering

Polish

Polish fowls are the Don Kings of the chicken world, because of their explosion of erect head feathers. They come in both bearded and non-bearded varieties and are referred to by many different names: Crested Fowl, Crested Dutch, Polands, Poland Fowls, Paduan, Padoue, and Patavinian. Ulisse Aldrovandi wrote about them in the sixteenth century, complete with illustration, and Charles Darwin classified them in the nineteenth century as fowls with topknots. Although their ancestry is debatable, many claim they originated in Padua, Italy. The Bearded Polish have been known in Holland since the sixteenth century. From there, they arrived in England in 1816 and were shown at the first poultry show in London in 1845.

The non-bearded varieties have a different ancestry. They are said to have originated in an area between Sweden and the Baltic nations of Estonia, Lithuania, and Latvia. This region was then in the hands of Poland, which may account for the breed's name. The birds were originally brought to Holland by Dutch seafarers in the Middle Ages. They are portrayed in many of the Old Dutch paintings, including one by Jan Monckhorst in 1657.

Large

Class: Continental (Polish)
Varieties: Non-Bearded White Crested Black, Non-Bearded Golden, Non-Bearded Silver, Non-Bearded White (1874); Bearded Golden, Bearded Silver, Bearded White, Bearded Buff Laced (1883); Non-Bearded Buff Laced (1938); Non-Bearded White Crested Blue (1963), Non-Bearded Black Crested White (1996)
Weight:
Cock: 6 pounds
Hen: 4.5 pounds
Cockerel: 5 pounds
Pullet: 4 pounds

Bantam

Class: All Other Combs Clean Legged Bantams
Varieties: Non-Bearded White (1883), Bearded White (1894), Bearded Buff Laced (1898), Non-Bearded Buff Laced (1904), Bearded Golden, Bearded Silver, Non-Bearded Golden, Non-Bearded Silver, Non-Bearded White Crested Black (1938), Non-Bearded White Crested Blue (1965)

Characteristics

Comb: Non-existent or small V-shaped
Tail: Large, the male's is carried at a 45-degree angle, the female's at 40 degrees.
Legs: Slate blue shanks and toes
Eggs: Small, white
Other: The Polish has a large, puffy crest or topknot that grows from a protuberance on the top of its skull. The male's crest falls down the nape in hackle-like feathers, while the female's is more rounded.

Status: Watch

Bearded Silver Polish

Bearded Golden, Bearded Buff Laced, and Bearded White Polish Bantams

Redcap

The Redcap gets its name from its large, symmetrically shaped rose comb that covers its head. It has been around since the fourteenth century, according to Martin Doyle's *Illustrated Book of Domestic Poultry* (1854), which cites a description of the fowl in Geoffrey Chaucer's "The Nonnes Preestes Tale." Poultry expert Reverend Edmund Saul Dixon claimed in the nineteenth century that Redcaps are a variety of the Hamburg. Others said they were a sub-variety of the Golden Spangled Hamburg. They are also called the Derbyshire Redcap or Yorkshire Redcap, having been bred in these areas of England. Now critically endangered, the fowl was popular in the late 1800s due to its prolific egg production and high meat quality. A hardy fowl and a good forager, the Redcap withstood cold winters and needed little attention, which added to its popularity around the world.

Redcaps (Courtesy of William A. Suys, Jr.)

Large

Class: English
Variety: Redcap (1888)
Weight:
Cock: 7.5 pounds
Hen: 6 pounds
Cockerel: 6 pounds
Pullet: 5 pounds

Bantam

Class: Rose Comb Clean Legged Bantams
Variety: Rose Comb (1960)

Characteristics

Comb: Rose. The male's is large and flashy. It is square and wide in the front and ends in a long and straight spike. The female's is about one-half the size of the male's.
Tail: Full, the male's is carried at an angle of 50 degrees, the female's at 45 degrees.
Legs: Leaden blue shanks and toes
Eggs: Small, white
Other: The plumage is a rich mahogany, with feathers ranging from black to red to rich brown.

Status: Critical

Rhode Island Red

The Rhode Island Red is an all-American breed. The town of Little Compton, Rhode Island, lays claim to its development in the 1800s, and the community has a ten-foot-high granite monument, dedicated in 1929, to memorialize the chicken. Rhode Island Reds were first exhibited in Boston in 1880, and the Rhode Island Club of American formed in 1898. A hardy dual-purpose bird, the Rhode Island Red soared in popularity in the first half of the twentieth century. Rhode Island Reds are a mixture of Red Malay Games, Leghorns, and Asian stock. They are the official state bird of Rhode Island.

Large

Class: American
Varieties: Single Comb (1904), Rose Comb (1905)
Weight:
Cock: 8.5 pounds
Hen: 6.5 pounds
Cockerel: 7.5 pounds
Pullet: 5.5 pounds

Bantam

Rhode Island Red bantams are in two classes.
Class: Rose Comb Clean Legged
Variety: Rose Comb (1952)
Class: Single Comb Clean Legged Other Than Game Bantams
Variety: Single Comb (1940)

Characteristics

Comb: Single or rose. The single is straight and upright with five even points, the center points larger. The rose is oval and terminates in a slightly drooping spike.
Tail: Medium sized, the male's is carried at 20 degrees, the female's at 10 degrees.
Legs: Rich yellow shanks and toes with a tinge of reddish horn
Eggs: Medium, brown

Status: Recovering

Single Comb Rhode Island Reds

Rhode Island White

The Rhode Island White has always lived in the shadow of its red counterpart, the Rhode Island Red, and never achieved an equal level of popularity. Rhode Islander J. Alonzo began developing the breed in 1888, mating Partridge Cochins, Rose Comb White Leghorns, and White Wyandottes. The new breed was presented to the public in 1903 and admitted to the *Standard* in 1922.

Large

Class: American
Variety: Rose Comb (1922)
Weight:
Cock: 8.5 pounds
Hen: 6.5 pounds
Cockerel: 7.5 pounds
Pullet: 5.5 pounds

Bantam

Class: Rose Comb Clean Legged Bantams
Variety: Rose Comb (1960)

Characteristics

Comb: Rose, oval-shaped and terminating in a slightly drooping spike
Tail: Medium sized, the male's is carried at 20 degrees, the female's at 10 degrees.
Legs: Yellow shanks and toes
Eggs: Large, brown

Status: Watch

Rose Comb Rhode Island Whites

Rose Comb Bantam

The Rose Comb Bantam is an old breed. It came from crosses of Black Hamburgs and "common bantams." Breeder Enoch Hutton developed the first two varieties, the Black and the White, in England. Rose Comb Bantams are unusual for their large, round white earlobes. The *American Standard* recognizes just three varieties, while the American Bantam Association recognizes twenty-six, including the Black Breasted Red, Blue Brassy, Blue Red, Brassy Back, Brown Red, Crele, Lemon Blue, Mottled, Red Pyle, Splash, and Wheaten.

Black and White Rose Comb Bantams

Bantam

Class: Rose Comb Clean Legged Bantams
Varieties: Black, White (1874), Blue (1960)
Weight:
Cock: 26 ounces
Hen: 22 ounces
Cockerel: 22 ounces
Pullet: 20 ounces

Characteristics

Comb: Rose, square in front and terminating in a long, pointed spike tilted slightly upward
Tail: Large and full, the male's is carried at a 40-degree angle, the female's at 35 degrees.
Legs: The Black and Blue varieties have slate-colored shanks and toes; the White has pinkish white.
Eggs: Tiny, white to cream
Other: Large, round white earlobes

Sebright

Silver and Golden Sebrights

The Sebright is a true bantam, having no large breed counterpart. One of the oldest English breeds, it was developed in the early 1800s by British agriculturalist Sir John Saunders Sebright. Sebright started developing the breed with a buff-colored bantam hen (probably a Nankin), a red cockerel without sickle feathers, and a small hen resembling a Golden Hamburg. He spent over thirty years perfecting the Sebright's laced feathers, beautiful coloration, and hen-feathering. In 1815, he organized the Sebright Bantam Club, the first poultry club specializing in an individual breed.

Bantam

Class: Rose Comb Clean Legged Bantams
Varieties: Silver, Golden (1874)
Weight:
Cock: 22 ounces
Hen: 20 ounces
Cockerel: 20 ounces
Pullet: 18 ounces

Characteristics

Comb: Rose, square in the front, terminating in a straight spike that extends almost horizontally
Tail: Full, carried at a 70-degree angle
Legs: Slate blue shanks and toes
Eggs: Tiny, white or creamy
Other: The rooster is hen-feathered, lacking sickle, hackle, and saddle feathers. Both varieties have feathers laced with lustrous black.

Status: Study

Shamo

The Shamo's ancestors arrived in Japan from Thailand in the seventeenth century. The word *Shamo* is a corruption of Siam, the former name of Thailand. The Shamo is considered a Japanese breed, having been developed there for cockfighting. Although Shamos are aggressive fighters, they are said to be tame and friendly. They were exported to Europe and the United States in the early twentieth century.

Large

Class: All Other Standard Breeds (Orientals)
Varieties: Black, Black Breasted Red, Dark (1981), Wheaten female (1996), Wheaten male (2000)
Weight:
Cock: 11 pounds
Hen: 7 pounds
Cockerel: 9 pounds
Pullet: 6 pounds

Bantam

Class: All Other Combs Clean Legged Bantams
Varieties: Wheaten, Black, Dark (1980)

Characteristics

Comb: Pea
Tail: Moderately long, drooping
Legs: Yellow shanks and toes, long and muscular legs
Eggs: Medium, pale brown
Other: The Shamo has large, piercing eyes; small or absent wattles; and short, hard feathers. It is an imposing bird, standing up to thirty inches tall, almost vertical on strong, muscular legs and with scanty feathering.

Status: Study

Black Shamos (Courtesy of Diane Jacky)

Sicilian Buttercup

Sicilian Buttercups

The Sicilian Buttercup's unique comb sits on its head like a regal tiara. The breed was developed on the Italian island of Sicily, though it is thought to have originated across the Mediterranean Sea in Tripoli, according to England's Rare Poultry Society. Its cup-shaped comb, which can hold rainwater, lends credence to the theory that the bird's ancestors evolved in the arid climate of North Africa. The assumption is that the chickens drank out of each others' combs.

Whatever their origin, Buttercups first arrived in America in 1835 and, around 1865, seemed to drop out of sight. Then, hatching eggs from a mining district of Sicily arrived in the late 1800s and became the origin of the current stock. In 1897, the breed was finally established, and the American Buttercup Club formed in 1912.

Large

Class: Mediterranean
Variety: Sicilian Buttercup (1918)
Weight:
Cock: 6.5 pounds
Hen: 5 pounds
Cockerel: 5.5 pounds
Pullet: 4 pounds

Bantam

Class: All Other Combs Clean Legged Bantams
Variety: Sicilian Buttercup (1960)

Characteristics

Comb: Buttercup—a complete circle of medium-sized points
Tail: Moderately large, the male's is carried at 45 degrees, the female's at 40 degrees.
Legs: Willow shanks and toes, yellow bottoms of the feet
Eggs: White, small
Other: The plumage of the male and female is markedly dissimilar. The male has rich reddish orange plumage and greenish black tail feathers. The female has golden buff hackles and buff body feathers covered with elongated, black diagonal spangles.

Status: Critical

Silkie

The Silkie bantam is named for its unique feathers that feel like silk. It was once called the Wooly Hen. In his thirteenth-century travel journals from China, Marco Polo describes a black chicken that has wool instead of feathers, raising speculation that the Silkie originated in China. Others trace it to India, Japan, and the Philippines. Aldrovandi in 1645 described them as having hair like a cat. During the sixteenth and seventeenth centuries, the breed found its way to Europe, where it was called Silk Fowl. It arrived in England in the mid-1800s, where breeders developed a strain with stronger feathers that retained the fluffy appearance. The Silkie has purple-blue skin, blue flesh, and black bones that, when ground up, the Chinese believe hold special powers. Silkie hens are maternal and make sweet mothers.

Bantam

Class: Feather Legged Bantams
Varieties: Bearded White and Non-Bearded White (1874); Bearded Black, Non-Bearded Black (1965); Bearded Blue, Non-Bearded Blue, Bearded Buff, Non-Bearded Buff, Bearded Gray, Non-Bearded Gray, Bearded Partridge, Non-Bearded Partridge (1996), Bearded Splash (2000)
Weight:
Cock: 36 ounces
Hen: 32 ounces
Cockerel: 32 ounces
Pullet: 28 ounces

Characteristics

Comb: Walnut, almost circular, and a deep mulberry or black
Tail: Round and short with shredded feathers
Legs: Slate blue shanks and toes, slightly feathered legs, five toes
Eggs: Small, light tint
Other: Silky, hair-like plumage and a large crest on the head. The beak is leaden blue, the face smooth, and the eyes black.

Non-Bearded Black and Non-Bearded White Silkies

Spanish

The White Faced Black Spanish is thought to be the oldest of the Mediterranean fowl, or at least the longest known. Columella described it in the first century AD as having large white ears. But early description of the fowl varied; some sources stated that the chickens had white faces, white earlobes, or red earlobes. Called "The Fowls of Castile" or "The Fowls of Seville," they came to Spain from the East via the Mediterranean Sea. From Spain, they made their way with Spanish merchants to the West Indies. The Spanish were bred in Spain and then developed in other countries. British records of the breed go back as far as 1572; it was an established breed there by 1815. It was shown at the first poultry show in Boston in 1849.

Large

Class: Mediterranean
Variety: White Faced Black (1874)
Weight:
Cock: 8 pounds
Hen: 6.5 pounds
Cockerel: 6.5 pounds
Pullet: 5.5 pounds

Bantam

Class: Single Comb Clean Legged Other Than Game Bantams
Variety: White Faced Black (1960)

Characteristics

Comb: Medium single with five points. The female's droops to one side.
Tail: Large and full, the male's is carried at a 45-degree angle, the female's at 40 degrees.
Legs: Dark slate shanks and toes
Eggs: Large, white
Other: The Spanish has a unique smooth, white, mask-like face. The earlobes are large and enamel white, and the wattles are long and thin.

Status: Critical

White Faced Black Spanish

Sultan

The Sultan originated in Constantinople (present-day Istanbul), where its popularity with Turkish rulers earned it the name "Sultan's Fowl." Sultans were kept in palace gardens as "moving flowers" and believed to enhance the gardens. They were exclusive to the ruling class and not permitted outside the palace. It was a great honor to be given a pair. The British tried for decades to obtain the breed, and it finally arrived in the United Kingdom in 1854. Similar in appearance to the Polish, the Sultan is sometimes called the Polish fowl of Turkey. They are the original chicken developed with vulture hocks, the breed's most distinctive characteristic.

White Sultans (Courtesy of Diane Jacky)

Large

Class: All Other Standard Breeds (Miscellaneous)
Variety: White (1874)
Weight:
Cock: 6 pounds
Hen: 4 pounds
Cockerel: 5 pounds
Pullet: 3.5 pounds

Bantam

Class: Feather Legged Bantams
Variety: White (1960)

Characteristics

Comb: Small V-shaped with two tiny horns
Tail: Large, carried at an angle of 60 degrees. The male's has an abundance of sickles and coverts.
Legs: Slate blue shanks and toes. They have five toes, the middle and outer of which are profusely feathered.
Eggs: Small, white
Other: The Sultan has a large rounded crest, a full beard and muff, and vulture hocks.

Status: Study

Sumatra

Black Sumatras

In its early history, the Black Sumatra inhabited its native island of Sumatra as well as other parts of Malaysia. Believed to descend directly from the red jungle fowl and possibly wild pheasants, the Sumatra was bred as a game fowl. Originally known as the Sumatra Game or the Black Pheasant, the breed arrived in Boston in 1847 direct from Sumatra and became popular among poultry fanciers. The fowl arrived in the British Isles a few decades later, where the British standard was created by English breeders Lewis Wright and Frederick Eaton. Like the Silkie, the Sumatra is black boned and has black or dark blue flesh. Some believe the Silkie and Sumatra interbred in their native lands.

Large

Class: All Other Standard Breeds (Orientals)
Variety: Black (1883)
Weight:
Cock: 5 pounds
Hen: 4 pounds
Cockerel: 4 pounds
Pullet: 3.5 pounds

Bantam

Class: All Other Combs Clean Legged Bantams
Varieties: Black (1960), Blue (1996)

Characteristics

Comb: Small pea
Tail: Long and flowing, carried horizontally and even drags on the ground
Legs: Black to dark willow shanks and toes, yellow bottoms of the feet
Eggs: Medium, white or tinted
Other: The Sumatra has multiple spurs, usually three or more on each leg. Single spurs would be a disqualification in exhibition. The plumage is greenish black and often considered to be the most lustrous of all the black breeds.

Status: Critical

Sussex

The county of Sussex, England, was a center of poultry in the 1800s and known for producing the best table fowl. The birds raised here became known as Sussex fowl. They had the body formation and white skin best suited for table poultry, but they also excelled as egg layers. A Sussex poultry club was formed in England in 1903. The original breed colors were brown, red, and speckled. Today's Light Sussex was developed from Brahmas, Cochins, and Silver Grey Dorkings. In England, the modern broiler chicken is a mix of the Sussex and Indian Game. ***Continued p. 132...***

Speckled Sussex

Large

Class: English
Varieties: Speckled, Red (1914), Light (1929)
Weight:
Cock: 9 pounds
Hen: 7 pounds
Cockerel: 7.5 pounds
Pullet: 6 pounds

Bantam

Class: Single Comb Clean Legged Other Than Game Bantams
Varieties: Light, Red, Speckled (1960)

Characteristics

Comb: Medium single with five well-defined points, the middle three larger
Tail: Medium length and well spread, the male's is carried at a 45-degree angle and the female's at 35 degrees.
Legs: Pinkish white shanks and toes
Eggs: Medium to large, creamy white to light brown
Other: The plumage of the Speckled variety is a beautiful mahogany, and each feather is tipped with white. The Red has lustrous mahogany red feathering and a greenish black tail. The Light has white plumage and lustrous green neck and tail feathers.

Status: Rare

Light Sussex

Welsummer

The Welsummer was developed in the Dutch village of Welsum at the beginning of the twentieth century. It is a mix of Partridge Cochins, Partridge Wyandottes, Partridge Leghorns, Barnevelders, Rhode Island Reds, and Croad Langshans. Welsummer hens lay dark brown eggs, some mottled with brown spots. When the breed was introduced in England in 1920 and the eggs were first sold, people thought them a hoax because the dark coloring can rub off.

Welsummers (Courtesy of William A. Suys, Jr.)

Large

Class: Continental (North European)
Variety: Welsummer (1991)
Weight:
Cock: 7 pounds
Hen: 6 pounds
Cockerel: 6 pounds
Pullet: 5 pounds

Bantam

Class: Single Comb Clean Legged Other Than Game Bantams
Partridge (1995)

Characteristics

Comb: Medium single with five distinct points
Tail: Full and large, carried at an angle of 60 degrees
Legs: Yellow shanks and toes
Eggs: Large, dark brown

Wyandotte

The Wyandotte was created in the United States in the 1800s, after many stages of development. The first pre-Wyandotte was called a Sebright Cochin and was developed in New York by crossing Silver Sebrights and White Cochins. These were then bred with Silver Spangled Hamburgs and Buff Cochins and called American Sebrights or Sebright Wyandottes. Michigan breeder L. Whittaker is credited with creating the original Silver Wyandotte in 1872 by mating a flock of Sebright Cochins with Dark Brahmas and Silver Spangled Hamburgs to create a new breed with clean legs and a rose comb. When this breed was admitted to the *Standard* in 1883, it was officially named Wyandotte, after the Native American nation. Other varieties were soon developed. The Golden was developed in 1880 in Wisconsin by crossing Partridge Cochins and Brown Leghorns with Silver Laced Wyandotte females. By 1905, White, Black, Buff, Partridge, Silver Penciled, and Columbian varieties were also admitted to the *Standard*. The breed's popularity stems from its dual purpose as a prolific egg layer of brown eggs and a good meat bird. ***Continued p. 136...***

Silver Laced Wyandottes

Silver Penciled Wyandottes

Columbian Wyandottes

Large

Class: American
Varieties: Silver Laced (1883), Golden Laced, White (1888), Black, Buff (1893), Partridge (1901), Silver Penciled (1902), Columbian (1905), Blue (1977)
Weight:
Cock: 8.5 pounds
Hen: 6.5 pounds
Cockerel: 7.5 pounds
Pullet: 5.5 pounds

Bantam

Class: Rose Comb Clean Legged Bantams
Varieties: Black, Buff, Columbian, Partridge, Silver Laced, Silver Penciled, White (1993), Golden Laced (1960), Buff Columbian (1965), Blue (1977)

Characteristics

Comb: Rose with a well-developed spike that follows the contour of the head
Tail: Short, the male's is carried at a 40-degree angle and the female's at 30 degrees.
Legs: Yellow shanks and toes
Eggs: Large, light to rich brown

Status: Recovering

White Wyandottes

Yokohama

The Yokohama is of Japanese origin and counts the Minohiki and the famous long-tailed Onagadori as its ancestors. The fowl not only boasts a long tail, but it also can live more than fifteen years, spending most of its life on high perches to protect its tail. The birds made their way to England through the Japanese port city of Yokohama, from which they earned their Western name. In Germany, the breed is known as the Phoenix. The only difference between the two is the comb, the Phoenix having a single and the Yokohama a walnut. Though the length of their tail does not match the Onagadori's, Yokohamas still have tails up to thirty-six inches long, which necessitate specially designed pens with high perches.

Large

Class: All Other Standard Breeds (Orientals)
Varieties: White, Red Shouldered (1981)
Weight:
Cock: 4.5 pounds
Hen: 3.5 pounds
Cockerel: 4 pounds
Pullet: 3 pounds

Bantam

Class: All Other Combs Clean Legged Bantams
Varieties: White, Red Shouldered (1981)

Characteristics

Comb: Rather small walnut, set just above the beak.
Tail: Long and carried horizontally, with many feathers dragging on the ground
Legs: Yellow shanks and toes
Eggs: Small, tinted light brown
Other: The Red Shouldered has a beautiful white head and tail. The body feathers range from red to salmon. Some of the feathers have white tips.

Status: Study

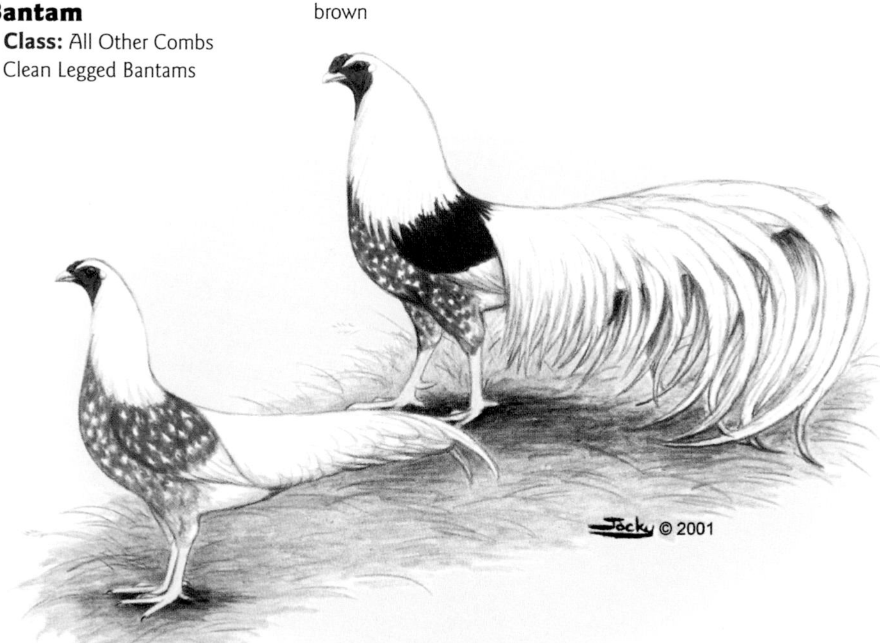

Red Shouldered Yokohamas (Courtesy of Diane Jacky)

Resources

Chicken Organizations

The American Poultry Association
Ric Ashcraft
5757 West Fork Rd.
Cincinnati, OH 45247
513-598-4337
www.amerpoultryassn.com

American Bantam Association
PO Box 127
Augusta, NJ 07822
973-383-6944
www.bantamclub.com

The American Livestock Breeds Conservancy
PO Box 477
Pittsboro, NC 27312
919-542-5704
www.albc-usa.org
albc@albc-usa.org

Society for the Preservation of Poultry Antiquities
Chuck Everett, Secretary
122 Magnolia Lane
Lugoff, SC 29078
crheverett@bellsouth.net
www.feathersite.com/Poultry/SPPA/SPPA.html

New England Heritage Breeds Conservancy, Inc.
244 Main St., Suite 2
Great Barrington, MA 01230
413-443-8356
contact@nehbc.org
www.nehbc.org

National Poultry Museum
The National Agricultural Center and Hall of Fame
630 Hall of Fame Drive
Bonner Springs, KS 66012
www.poultryscience.org/psapub/pmuseum

Clubs

Listed alphabetically by chicken breed.

Ameraucana Breeders Club
John Blehm, Secretary/Treasurer
4499 Lange
Birch Run, MI 48415
989-777-1234
www.ameraucana.org

Arauanca Club of America
Nancy Utterback
11683 North 600 West
Frankton, IN 46044
Araucana21@hotmail.com
www.araucana.freehosting.net

Belgian Bearded d'Anver Club of America
Greg Romer, Secretary
7237 California Lane
Okeana, OH 45053
513-738-4012
gsromer@msn.com
www.danverclub.com

Belgian d'Uccle & Booted Bantam Club
Kimberly Theodore, Secretary/Treasurer
9N 100 Percy Rd.
Maple Park, IL 60151
Theodk@aol.com
www.belgianduccle.org

American Brahma Club
Sandy Kavanaugh, Secretary/Treasurer
216 Meadowbrook Rd.
Richmond, KY 40475
859-369-7244
henshaven@iclub.org
http://groups.msn.com/American-BrahmaClub

American Buttercup Club
Julie Cieslak, Secretary/Treasurer
7257 W. 48 Rd.
Cadillac, MI 49601
231-862-3671
americanbuttercup@yahoo.com
www.geocities.com/american-buttercupclub

Cochins International
Dennis Wollard, Secretary/Treasurer
1913 Beau Bassin Rd.
Carencro, LA 70520
318-896-5442
wollard@louisiana.edu
www.geocities.com/cochin-man2005/

International Cornish Breeders Association
Billy Grimes, Secretary
PO Box 373
Leonard, TX 75452
903-587-2950
billyg@fanninelectric.com
www.inter.cornish.org

Dominique Club of America
Tracey Allen, Secretary/Treasurer
113 Ash Swamp Rd.
Scarborough, ME 04074
domchickens@earthlink.com
www.dominiquechickens.org

Dorking Club of America
Phillip Bartz, Secretary
1296 Perbix Rd., Route 1
Chapin, IL 62628
217-243-9229
Rooster688@hotmail.com

American Dutch Bantam Society
Ric Ashcraft, President
5757 West Fork Rd.
Cincinnati, OH 45247
AshleyPines@aol.com
www.dutchbantamclub.com

Faverolles Fanciers of America
Dick Boulanger
69 Perry St.
Douglas, MA 01516
508-476-2691
FAVEROLLES@prodigy.net
www.faverollesfanciers.org

National Frizzle Bantam Club
Loretta Schmidgall, President
Route 2, Box 785
Thomasville, PA 17364

National Frizzle Club of America
Glenda Heywood,
Secretary/Treasurer
207 13th Ave
Brookings, SD 57006
864-855-0140
frizzlebird@yahoo.com
www.g-kexoticfarms.com

American Game Bantam Club
Lacy Greer, Secretary/Treasurer
19006 E. Karsten Dr.
Queen Creek, AZ 85242
http://groups.msn.com/agbc

Old English Game Bantam Club of America
Sharon Garrison,
Secretary/Treasurer
316 Sullivan Rd.
Simpsonville, SC 29680
864-299-0901
syg4138@aol.com
www.bantychicken.com/OEGBCA

The Modern Game Bantam Club of America
Lee A. Traver, Secretary/Treasurer
4134 NY 43
Wynantskill, NY 12198

North American Hamburg Society
Mary Hoyt, Secretary/Treasurer
9365 N Santa Margarita Rd.
Atascadero, CA 93422
805-466-3185
hoytwoodacre@charter.net
www.geocities.com/north-americanhamburgsociety

Japanese Bantam Breeders Association
John deSaavedra, President
5899 Blacks Rd.
Pataskala, OH 43062
http://home.columbus.rr.com/jbba/JBBA.html
johnde@ameritech.net

North American Java Club
1408 Mason Bay Rd.
Jonesport, ME 04649
207-497-3431
pamhlm@raccoon.com
www.northamericanjavaclub.com

National Jersey Giant Club
Robert Vaughn, Secretary
28143 Country Road 4
Pequot Lakes, MN 56472
218-562-4067

American Langshan Club
Forest Beaufford
Route 5, Box 75
Claremore, OK 74017
918-341-2238

American Brown Leghorn Club
Dennis Pearce,
Secretary/Treasurer
PO Box 602
Stanwood, WA 98292
360-629-3356
www.the-coop.org/leghorn/ablc1

North American Marans Club
Cari Shafer
PO Box 17142
Hattiesburg, MS 39404
www.maransclub.com

National Naked Neck Breeders Society
Ed Haworth, Secretary
Route 1, Box 322
Tahlequah, OK 74464

New Hampshire Breeders Club of America
Edgar Mongold, Secretary
918 Stuckey Rd.
Washington Court House, OH 43160
740-333-5080
edgar@mongold1.com
www.mongold1.com/nhbca

United Orpington Club
Richard Andree,
Secretary/Treasurer
105 Johnson St. NE
Brownsdale, MN 55918
507-567-2157
rmandree@mchsi.com
www.geocities.com/srp18407/UOC.html

Plymouth Rock Fancier's Club of America
1988 Cook Rd.
Lucasville, OH 45648
614-259-2852
www.crohio.com/rockclub

Crested Fowl Fanciers Association
Rick Porr
72 Springer Lane
New Cumberland, PA 17070
717-774-1926

Polish Breeders' Club
Jim Parker, Secretary/Treasurer
RR #6, 3232 Schooler Road
Cridersville, OH 45806
polishman@watchtv.net
http://groups.msn.com/Crested-BreedersClub

Rhode Island Red Club of America
William Post, Secretary
209 County Highway 17
New Berlin, NY 13411
postwill1@yahoo.com
John Klimes
jklimes@agri.state.id.us
www.crohio.com/redclub

Rosecomb Bantam Federation
Fran Curtis, Secretary/Treasurer
PO Box 109
Thomaston, ME 04861
207-354-0711
fjcurtis@midcoast.com
www.rosecomb.com/federation

Russian Orloff Club of America
Erin Traverse
PO Box 69
Poultney, VT 05764
802-287-2007
www.feathersite.com/Poultry/Clubs/Orloff/OrlClub

Sebright Club of America
Mary Ann Bonds
PO Box 136
Ila, GA 30647
706-789-2869

U.S. Serama Club
Brian Sparks, Club Leader
Markesan, WI 53946
www.serama.us
brian@w3gate.com

National Serama Club and Registry
Anthony De Piante
8501 Flowe Farm Rd.
Concord, NC 28025
www.seramaclub.com

Serama Council of North America
Jerry Schexnayder
PO Box 159
Vacherie, LA 70090
www.seramacouncilofnorth-america.com

The Oriental Game Breeders' Association
Eve Bundy
PO Box 100
Creston, CA 93432
805-237-1010

American Silkie Bantam Club
Sheila Gordon, Secretary/Treasurer
276 E Palo Verde Ave.
Palm Springs, CA 92264
760-320-5960
sgordonwindsor@earthlink.net
www.americansilkiebantamclub.org

American Sumatra Association
Richard Schock
3036 Woodruff Rd.
Boonville, NC 27011
336-367-5882
rschock@yadtel.net

Sussex Club of America
John Wolf
7959 Huff
Acton, IN 46259
317-862-6515
KWolf77434@aol.com

Wyandotte Breeders of America
Dave Lefeber, Secretary/Treasurer
8648 Irish Ridge Rd.
Cassville, WI 53806
608-725-2179
dottestuff@yahoo.com
www.crohio.com/wyan

Silver Laced Wyandotte Club
Robert Coulter
Route 2, Box 151
Owatonna, MN 55060
507-451-4274
rcoulter@mnic.net

Chicken Periodicals

Fancy Fowl
The Publishing House, Station Rd.
Framlingham, Suffolk
England IP13 9EE
01728 622030
www.fancyfowl.com
fancyfowl@my-deja.com

Feather Fancier Newspaper
Paul Monteith, Owner/Editor
5739 Telfer Rd
Sarnia, Ontario N7T 7H2
Canada
519-542-6859
featherfancier@ebtech.net
featherfancier.on.ca

National Poultry News
Glenda Heywood
207 13th Ave.
Brookings, SD 57006
605-692-1029
www.nationalpoultrynews.com
nationalpoultrynews@yahoo.com

Poultry Press
PO Box 542
Connersville, IN 47331
765-827-0932
www.poultrypress.com
info@poultrypress.com
poultryp@si-net.com

Show Bird Journal
Shelia Clanton, Editor
29971 County Road 8
Florence, AL 35634
256-757-9324
showbird@vtechworld.com

Other Important Websites

AvianWeb: All About Birds
www.avianweb.com/chicken.htm

FeatherSite
www.feathersite.com

The ICYouSee Handy-Dandy Chicken Chart
www.ithaca.edu/staff/jhenderson/chooks/chooks.html

Poultry Breeds, Department of Animal Science, Oklahoma State University
www.ansi.okstate.edu/poultry

The Poultry Club of Great Britain
www.poultryclub.org

Sand Hill Preservation Center
www.sandhillpreservation.com

Urban Programs Resource Network, University of Illinois Extension
www.urbanext.uiuc.edu

Bibliography

Aldrovandi, Ulisse. *Aldrovandi on Chickens*. Translated by L. R. Lind. Norman: University of Oklahoma Press, 1963.

American Poultry Journal. 7th World's Poultry Congress, July and August, 1939.

American Standard of Excellence. Buffalo, NY: American Poultry Association, 1874.

American Standard of Perfection. Mendon, MA: American Poultry Association, 1905.

American Standard of Perfection. Mendon, MA: American Poultry Association, 1998.

Bement, C. N. *The American Poulterer's Companion.* New York: Harper & Brothers, 1867.

Bender, Marjorie E. F., Donald E. Bixby, and Robert O. Hawes. *Counting Our Chickens: Identifying Breeds in Danger of Extinction.* Pittsboro, NC: American Livestock Breeds Conservancy, 2004.

British Poultry Standards. Edited by Victoria Roberts. 5th ed. England: Blackwell Science, 1997.

Gordon, John Steele. "The Chicken Story." *American Heritage*, September 1996: 52-67.

Howard, George E. *The American Fancier's Poultry Book*. Washington, DC. 1898.

Jull, M. A. "The Races of Domestic Fowl." *National Geographic*, April 1927: 379-452.

Limburg, Peter R. *Chickens, Chickens, Chickens*. Nashville, NY: Thomas Nelson, 1975.

Luttmann, Rick and Gail. *Chickens in Your Backyard.* Emmaus, PA: Rodale Press, 1976.

McGrew, Thomas F. *The Book of Poultry*. New York: Thomas Nelson and Sons, 1926.

Meall, L. A. *Moubray's Treatise on Domestic and Ornamental Poultry*. London: Arthur Hall, Virtue and Co., 1854.

Platt, Frank L. "All Breeds of Poultry." *American Poultry Journal*, 1925.

Porter, Valerie. *Domestic and Ornamental Fowl.* London: Pelham Books, 1989.

Roberts, Victoria. *Poultry for Anyone*. Suffolk, UK: Whittet Books, 1998.

Skinner, John L. *Establishing the Breeds*. Madison: University of Wisconsin.

—. *Bantams*. Wisconsin: North Central Regional Publication Extension 209.

Stern, Alice. *Poultry and Poultry-Keeping*. London: Merehurst Press, 1988.

Thear, Katie. *Free-Range Poultry*: Ipswitch, UK, 1997.

Van Hoesen, R. W. *History of the Anconas*. Franklinville, NY: Ancona World, 1915.

This 1849 Currier & Ives lithograph depicts "the game cock in full feather." (Courtesy of the Library of Congress)

Index

Aldrovandi, Ulisse, 17, 18, 95
Ameraucanas, 58, 59, 66
American Livestock Breeds Conservancy, 62, 63, 89
American Poultry Association, 25, 61
American Standard of Perfection, 13, 19, 20, 25, 30, 32, 33, 35, 61, 63–65, 68, 69, 72, 74, 85, 92, 94, 102, 110, 114, 115, 122, 123, 134
Anconas, 30, 31, 46, 58, 59, 67
Andalusians, 31, 32, 46, 58, 59, 68
Araucanas, 28, 33, 58, 63, 69
Aseels, 28, 31, 32, 35, 58, 64, 70
Australorps, 38, 46, 58, 59, 63, 71, 97
bantams, 14, 15, 32, 34, 53, 61, 63, 64
Barnvelders, 58, 72
behavior, 37–41
Belgian Bearded d'Anvers, 30, 32, 34, 59, 73, 74
Belgian Bearded d'Uccles, 34, 59, 64, 74
Booted Bantams, 59, 75
Brabanters, 21
Brahmas, 27, 28, 31, 33, 46, 58, 59, 72, 76, 77. 93, 101, 131
Buckeyes, 28, 34, 58, 78
Burnham, George P., 25
Campines, 19, 30, 31, 58, 59, 79
Catalanas, 33, 58, 59, 80
Chaam Fowls, 20
Chanteclers, 28, 34, 58, 81
chicks, 45–47, 53
China, chickens in, 15
Cochins, 27, 30–34, 41, 46, 58, 59, 64, 72, 78, 82–84, 105, 117, 122, 131, 133, 134
colors: egg, 45
colors: feathers, 27, 32–35
combs, 28, 29, 51, 64
Cornish, 28, 30, 31, 33, 39, 58, 64, 81, 85, 88
Crevecoeurs, 29, 45, 58, 86. 93, 98, 102
Cubalayas, 28, 32, 58, 64, 87
Delawares, 9, 47, 58, 59, 88
Dominiques, 30, 31, 58, 59, 89, 117
Dorkings, 30, 34–37, 49, 58, 59, 90, 91. 93, 104, 117, 131
Drenthe Fowls, 21
Dutch Bantams, 32, 33, 59, 92, 99
eggs, 38, 40–46, 65
England, 19
Etruscans, 17
Faverolles, 34, 58, 59, 93
Fayoumis, 16
feathers, 27, 30, 31, 50, 51
France, chickens of, 19, 22
Frizzles, 26, 27, 31, 41, 58, 59, 94
Games, *See* Modern Games, Old English Games
Hamburgs, 30, 31, 58, 59, 95, 96, 124, 134
Holland, 30, 58, 59, 97
Houdans, 29, 30, 58. 93, 98
Iran, chickens of, 16, 17
Japan, chickens of, 15, 16
Japanese Bantams, 30, 33, 35, 59, 61, 99
Javas, 30, 35, 58, 59, 100, 117
Jersey Giants, 46, 58, 59, 101
jungle fowls, 13, 14, 117, 130
Kedu Cemanis, 14
La Flèche, 29, 58, 102
Lakenvelders, 58, 59, 103
Lamonas, 58, 59, 97, 104
Langshans, 32, 58, 59, 101, 105, 115, 133
Leghorns, 30–35, 46, 47, 58, 59, 81, 97, 104, 106, 121, 122, 133, 134
Malays, 29, 32, 58, 64, 72, 107, 117, 121
Malines, 19
Marans, 22
Minorcas, 29, 33, 58, 59, 102, 108, 109, 115
Modern Games, 32–34, 58, 59, 110, 111
Naked Necks, 34, 58, 59, 112
New Hampshires, 47, 58, 88, 97, 113
Old English Games, 31–34, 35, 58, 59, 110, 114
Orpingtons, 32, 33, 58, 59, 63, 101, 115
pecking order, 39, 40
Phoenix, 35, 58, 59, 116
Plymouth Rocks, 30–35, 39, 46, 47, 58, 59, 63, 78, 81, 104, 115, 117
Polish, 29–31, 46, 53, 58. 93, 98, 102, 118, 119
Redcaps, 29, 58, 59, 120
Rhode Island Reds, 29, 34, 37, 46, 47, 58, 59, 63, 81, 97, 113, 121, 133
Rhode Island Whites, 29, 58, 59, 63, 64, 122
Roman Empire, 17
Rose Comb Bantams, 29, 59, 123
Russian Orloffs, 17
Sebrights, 28, 30, 31, 34, 59, 61, 124, 134
Seramas, 14, 15
Shamos, 35, 58, 59, 125
Sicilian Buttercups, 28, 31, 58, 126
Silkies, 28, 29, 33, 34, 41, 59, 61, 64, 127, 130
Society for the Preservation of Poultry Antiquities, 62
Spanish, White Faced Black, 58, 59, 128
Sultans, 29, 58, 59, 129
Sumatras, 28, 28, 32, 34, 58, 59, 64, 130
Sussex, 33, 34, 58, 59, 131, 132
tails, 27, 57, 64
Thailand, chickens of, 13, 16
wattles, 27, 57
Welsummers, 58, 59, 133
wings, 27
Wyandottes, 29–31, 33–35, 58, 59, 63, 81, 122, 133–136
Yokohamas, 29, 58, 59, 137

The New Hampshire inherited its rich red plumage from its forbear, the Rhode Island Red. (Photograph © Alan and Sandy Carey)